The Isolated State in Relation to
Agriculture and Political Economy

Portrait of Johann Heinrich von Thünen by Wilhelm Ternite, 1840 (Thünen-Museum-Tellow).

The Isolated State in Relation to Agriculture and Political Economy

Part III: Principles for the Determination of Rent, the Most Advantageous Rotation Period and the Value of Stands of Varying Age in Pinewoods

Johann Heinrich von Thünen

Translated from the German by
Keith Tribe

Edited by
Ulrich van Suntum
On behalf of Thünengesellschaft e.V., Tellow

First published in German 1863

English translation published 2009 by
PALGRAVE MACMILLAN

Palgrave Macmillan in the UK is an imprint of Macmillan Publishers Limited, registered in England, company number 785998, of Houndmills, Basingstoke, Hampshire RG21 6XS.

Palgrave Macmillan in the US is a division of St Martin's Press LLC, 175 Fifth Avenue, New York, NY 10010.

Palgrave Macmillan is the global academic imprint of the above companies and has companies and representatives throughout the world.

Palgrave® and Macmillan® are registered trademarks in the United States, the United Kingdom, Europe and other countries.

ISBN-13: 978-0-230-22251-9 hardback
ISBN-10: 0-230-22251-X hardback

This book is printed on paper suitable for recycling and made from fully managed and sustained forest sources. Logging, pulping and manufacturing processes are expected to conform to the environmental regulations of the country of origin.

A catalogue record for this book is available from the British Library.

A catalog record for this book is available from the Library of Congress.

10 9 8 7 6 5 4 3 2 1
18 17 16 15 14 13 12 11 10 09

Printed and bound in Great Britain by
CPI Antony Rowe, Chippenham and Eastbourne

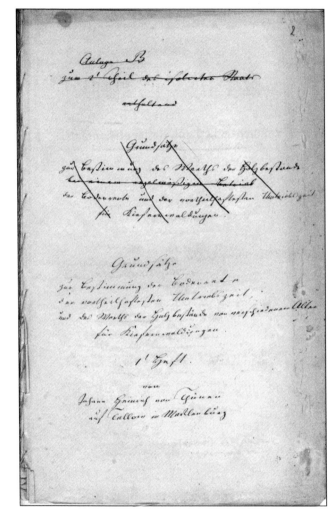

Frontpage, written by Johann Heinrich von Thünen, deposited at the Rostock University, **U.A.R. TA F I 13.**

Der isolirte Staat

in Beziehung auf

Landwirthschaft und Nationalökonomie.

Von

Johann Heinrich von Thünen.

Dritte Auflage,

herausgegeben von

H. Schumacher-Zarchlin.

Dritter Theil.

Grundsätze zur Bestimmung der Bodenrente, der vortheilhaftesten Umtriebszeit und des Werths der Holzbestände von verschiedenem Alter für Kieferwaldungen.

~~~~~~~~~~~~~~~~~~~~

Berlin.

Verlag von Wiegandt, Hempel & Parey.

Verlagsbuchhandlung für Landwirthschaft, Gartenbau und Forstwesen.

1 8 7 5.

The Isolated State, 3rd edition at the Thünen-Museum-Tellow; frontpage of the 3rd part.

Contents

Foreword to the English Translation, by Fritz Tack   x

Preface, by Paul A. Samuelson   xii

Acknowledgements   xv

*Johann Heinrich von Thünen's Contribution to
Forestry Economics: A Brief Introduction to Part III of
the Isolated State, by Ulrich van Suntum*   xvi

Section One   1

§ 1. Timber yield   1
§ 2. Timber value   2
§ 3. The determination of the value of
    pinewood stock of a given age   5
§ 4. The determination of the value of
    timber stocks of differing rotation periods   7
§ 5. The rent of the woodland soil   9
§ 6. Calculations related to the rotation period   12
§ 7. Thinning out   18
§ 8. The rent of woodland in comparison with
    ground rent   26
§ 9. Underwood or brushwood   29
§ 10. The influence of the yield in value of thinnings
    on the most advantageous rotation period and
    on ground rent   32
§ 11. The thinning methods employed by
    Chief Forester Nagel   34
§ 12. Ground rent and rotation period from
    pollination, if thinning removes half the
    wooded growth   39

§ 13. Comparison of the two methods, in which
    thinning removes   41
A. One-third of annual timber growth
B. Half of this growth
Concerning the part of the volume, which is
    captured by thinning   41
Concerning the rent of woodland   42
Concerning the ground rent   44
Concerning the rotation period   46

Section Two   48

§ 14. How great does the space around each tree
    need to be in relation to its diameter if
    the annual growth in value of the
    entire woodland is to achieve a maximum?   48
§ 15. The growth of individual trees in
    diameter and in physical volume   51
§ 16. Calculation of the portion of the growth to be
    removed in thinning   53
§ 17. Criticism   64
§ 18. Ground rent and the most advantageous
    rotation period, if only ⅓ of the growth is left in
    the remaining stock   74

Section Three   78

§ 19. What is the relationship between the
    growth of the tree and the space which
    each tree is given?   78
§ 20. Calculation of total growth   90
§ 21. Ground rent and most advantageous rotation
    period for different degrees of interval
    between trees   95
§ 22. What part of the timber reserve is at
    different ages removed in decennial thinnings?   100
§ 23. Deviation of the results of
    our calculation from reality   102

§ 24. Comparison of the ground rent of
    woodland with that of arable land   105
§ 25. Applications   110

Continuation   114

**Epilogue: A Chronicle of Editing the
*Isolated State***   134

Glossary   137

Foreword to the English Translation

The small Mecklenburg village of Tellow, near Teterow, is a place that makes connections. Connections with the past, to future possibilities for the permanent development of the countryside, and to the wider world. But no gold or oil has been discovered here, and there is not even a hot spring. This small village has become well known because it was here, 200 years ago, that Johann Heinrich von Thünen, economist, model farmer, and humanist, forged a connection from his own estate to the wider world beyond.

In 1972 Rolf-Peter Bartz, then a teacher but now Director of the Tellow Museum, created with some of his pupils a national memorial to Thünen, now known as the 'Thünen Museum, Tellow'. Buildings and grounds of the former estate were rescued from dilapidation and collapse and opened to the public. In 1990, following German Unification, the Thünen Society (Thünengesellschaft e.V.) was formed at Tellow – out of the 'classical soil of German political economy' in the words of one of the most well-known Thünen scholars, Professor Asmus Petersen – with members from the region and beyond. Today the Society has an international membership, not only including German and European members, but also Americans and Japanese.

The official aim of the Thünen Society is to 'promote the study and dissemination of Thünen's wealth of ideas in today's world. This includes his works as an economist, as a farmer and thinker on agrarian policy, as well as his ideas and practical activity in search of a solution to the social question.' This objective will be met with the publication of the present work. The first two parts of Thünen's principal work were published in English in 1960[1] and 1966;[2] but until now little international note has been taken of Part III, for lack of a translation. The members of the Thünen Society are very pleased to remedy this gap with this translation by Keith Tribe. They are convinced that this will open up Thünen's work to further study and discussion.

My cordial thanks are owed to all members of the Thünen Society who contributed to the completion of this ambitious project. In particular I thank Reinhard Schwarze, who has not only initiated and funded the project but also rendered outstanding services to the editing of Thünen`s text with its many ancient and technical terms.

I hope that the work receives the wide attention it deserves, and express my thanks to those members of our Society who have assisted in this demanding endeavour.

Professor Fritz Tack, Rostock
University of Rostock,
Chairman of the Thünen Society,
Tellow/Mecklenburg

---

1. Chicago, edited by Bernard W. Dempsey.
2. London, translated by Carla M. Wartenberg and edited by Peter Hall.

Preface
Thünen: An Economist ahead of His Times

Both Alfred Marshall and Joseph Schumpeter paid homage to Johann Heinrich von Thünen as a prophet way ahead of his own times. Thünen was not only a theorist. He deduced the principles best suited to run his agricultural estate in the far east of Germany, hundreds of miles away from the epicentre of the Scottish Enlightenment where lived such as David Hume, Adam Smith, John Stuart and James Mill. Like France's Augustin Cournot (1838), Thünen anticipated the kind of mathematics later employed by Stanley Jevons, Léon Walras, Francis Edgeworth and Vilfredo Pareto.

While deserving notice and praise from his contemporaries, Thünen's work never really received the recognition it deserved. His was not the luck of Austrian German language writers (such as Böhm-Bawerk) whose works were promptly translated into English in the period when Anglo-Saxon economic literature dominated frontier discussions.

Admirers of Thünen have differed in what interested them most. Thünen's picturesque circular zones of specialization around a city eclipsed his later (1850s) anticipation of post-1890 marginal productivity theory. As late as 1983, coinciding with Thünen's 200th birthday, Samuelson (1983) for the first time puzzled out the general equilibrium system implicit in Thünen's natural wage as the geometric mean between workers' average productivity and their marginal productivity.

Only in the present millennium has Thünen's late work on forestry optimization been translated into English and become the object of renewed discussion. In World War II, when Allied physicists stumbled onto operation research, analyses like Thünen's came into fashion. By the mid-nineteenth century, when he was deep into middle age, in his solitary study Thünen deployed not only calculus but also Cardan's formula for finding the roots of a cubic equation. Applied maths work is never done. Whether to fell trees at 30 years or 90 years is sensitive to assumptions about numerical parameters and on the scope of included variables.

Not all great scholars are great human beings. At George Washington's death, he freed his slaves. Thomas Jefferson never did free his own slaves. This was not because he approved of that institution, but rather because of his own spendthrift ways. Thünen seems to have been one of those rare species of *homo economicus* who did indulge in random acts of altruism. And non-random ones, too.

The importance of German science was recognized in the nineteenth and early twentieth centuries: Gauss, Humboldt, Einstein and beyond. My generation of economists knew too little about the sophisticated analytical economist literature – Austrian and non-Austrian – independently generated on the continent. David Ricardo had his Robert Malthus and Mill. Thünen perforce remained a pre-Gossen loner.

It is fitting then that 225 years after his birth, Johann Heinrich von Thünen still speaks to us. It is appropriate to conclude this tribute with hot-off-the-press news about Thünen's capital theory, maybe the jewel in his crown.

Samuelson (1983) sketched Thünen's scenario in which, for example, final consumable corn gets produced by direct land and direct labour working with a produced capital input, K. That K, in Thünen's simplest mid-nineteenth century model, could be produced by labour alone. At my suggestion, a perceptive young Finn from the Massachusetts Institute of Technology and Harvard, Erkko Etula (2008), analysed definitely the more complicated Thünen case where K as well as corn gets produced by K and labour. Both in terms of post-1890 Clark–Wicksteed differentiable marginal productivities and (Samuelson and Etula, 2006) completion of the limited substitutability technology of Sraffa (1960) type, a cogent Thünen-inspired answer can finally be given to David Ricardo's fundamental query: how does a competitive regime answer the puzzle of how society's total harvest gets divided among (1) the rent to landowner (2) the wages of laborers and (3) the interest yield of (vectoral!) K's.

Yes. Science does advance funeral by funeral. And science does advance also birth by birth. Thünen's brainchildren live on and proliferate.

Paul A. Samuelson
Massachusetts Institute of Technology

References

Cournot, Antoine-Augustin (1838) *Recherches sur les Principles mathématiques de la Théorie des Richesses*. Paris: Hachette.

Etula, Erkko (2008) 'The Two Sector von Thünen Original Marginal Productivity Model of Capital; and beyond', *Metroeconomica* 59(1), pp. 85–104.

Samuelson, Paul A. (1983) 'Thünen at Two Hundred', *Journal of Economic Literature*, pp. 148–1488.

Samuelson, Paul A. and Erkko Etula (2006) 'Complete Work-Up of the One-Sector Scalar-Capital Theory of Interest Rate: Third Installment Auditing Sraffa's Never-Completed "Critique of Modern Economic Theory"', in *Japan and the World Economy*, pp. 331–56.

Sraffa, Piero (1960) *Production of Commodities by Means of Commodities: Preclude to a Critique of Economic Theory*. Cambridge: Cambridge University Press.

Acknowledgements

The English translation was generously supported by Reinhard Schwarze, Hamburg, member of the German Thünengesellschaft e.V., Tellow.

Johann Heinrich von Thünen's Contribution to Forestry Economics: A Brief Introduction to Part III of the *Isolated State*

While issues relating to forestry economics are raised only in the context of location theory in the first two parts of von Thünen's *Isolated State*, in this Part III they become of central importance. Thünen proceeds as before, starting out with extremely simplified assumptions and, as the investigation proceeds, taking ever greater account of reality. At its heart lies the question of the optimal rotation period of a wood.[3] This is a capital–theoretic problem that many later economists also studied – coming to surprisingly varied conclusions, as Paul Samuelson shows in his 1976 paper. Debate on this issue has still not reached a final conclusion, and Thünen's early approach to the problem compares very well with many later attempts (Manz 1986; van Suntum 1995; Helmedag 2002). The results established by Thünen were revolutionary in comparison with the forestry practices of the time – demonstrating that the relatively short rotation periods usually employed were very inefficient both from the point of view of the individual enterprise and for the wider economy (see §§ 24, 25). According to Thünen, this productive inefficiency accounted for the prevailing high level of timber imports to Mecklenburg, at the same time quite casually articulating an entirely

---

3. *Umtriebszeit* – the period from planting to harvesting of stands of timber.

correct account of the principle of comparative advantage (§ 25A).

A reader mainly interested in the basic principles of Thünen's solution and less concerned with the details of forestry economics need here read only Paragraphs 1 to 6. Thünen's procedure in these paragraphs is very like that of the German forester Martin Faustmann, who in 1849 published an account of the problem that remains the model for all later attempted solutions.[4] Thünen died in 1850, and Part III of *The Isolated State* was edited and published by Thünen's student and biographer Hermann Schumacher only in 1863, so that Thünen and Faustmann were each entirely unaware of the other's work. Like Faustmann, Thünen works with a model of the so-called synchronized wood, in which timber of all ages stands together on equal plots of land; and he also adopts Faustmann's approach in seeking the number of age classes that will maximize the land rent (and not for example the internal rate of return). To do this he subtracts from the proceeds of the sale of the oldest age class the annual planting costs for the youngest generation of trees, together with the return on the capital represented by the entire stand of timber. The resulting figure, divided by the number of age classes, gives the land rent per *Waldmorgen*, as Thünen calls the land required by each age class (§ 5). Differentiating this magnitude determines the maximum of land rent for a typical Mecklenburg pinewood, and defines the optimal rotation period for the wood as 42 years (§ 6).

This approach is in principle entirely correct, but leads to a shorter rotation period than that given by Faustmann's very similar procedure. The problem here relates to the calculation of the capital stock represented by the wood

---

4. For the contributions of some English prerunners see Scorgie and Kennedy (1996).

(van Suntum 1995; Nellinger 2002). Thünen equates this with the sum of the proceeds from the sale of each age class on the simultaneous sale of the entire synchronized wood (§ 3). But this value is too low if, instead of a one-off sale, the wood is allowed to gradually mature without replacement planting. The difference can be seen by considering the five youngest age classes. Thünen assigns a value of zero to these, since new plantings take five years to establish themselves as trees and so during this period have no saleable value (§ 1). But at the same time these new classes embody costs arising from their planting and the land on which they stand, costs that are later paid off by the sale of mature timber and which therefore must be included in the cost of capital. It is mainly here that Thünen diverges from Faustmann, whose procedure, using Thünen's assumptions, gives the much longer rotation period of 75 years.

In the course of his study Thünen takes particular account of the influence on the optimal rotation period of regular thinning-out of the wood. Thinning in this way costs money, but produces saleable timber while also giving the remaining woodland more room to grow. Depending on the nature and frequency of such thinning-out the optimal rotation period can increase quite markedly, to a maximum of 92 years (§ 21). Of course, the actual figures used by Thünen are today superseded. But the algebraic representation of all his calculations (§§ 19–21) lends this account, like his location theory, a timeless value for all later work in the area.

Thünen is conscious of the fact that his idealized forestry model cannot be found in reality. He describes his methodology in § 17: 'These investigations directed to the natural laws governing forestry relate to practical forestry as pure geometry relates to applied.' But he does not simply leave it at that; in § 23 he takes express account of potential objections to his model, such as forest fires, storm damage, and theft, finally deducting 50% from the timber returns. In the final section of his work ('Continuation') he raises further issues, in particular the availability of especially hard timber. This requires a tree much older than 92 years and would not therefore be part of the yield from such a rotation period. Thünen shows that with the increasing dispersal of trees resulting from progressive thinnings the optimal rotation period rises to 140 years, and that this remains profitable because of the advanced age of the timber (Section seven of the 'Continuation').

Part III of *The Isolated State* remains unfinished. We do not even know whether Thünen would have agreed to its posthumous publication, especially of the fragments included in the final part. Some parts of the text go back to 1828 and were taken up by Thünen much later, as is clear in § 17. Thünen certainly could not have completed all the tasks laid out in the opening pages of the 'Continuation'. But these are more by way of empirical details which, while relevant to contemporary forestry practice, do not affect the substance of his study. Thünen's connection of the theory of investment to practical forestry will always be a milestone in the literature of economics.

Ulrich van Suntum,
University of Münster

References

Faustmann, Martin (1849) 'Berechnung des Werthes, welchen Waldboden, sowie noch nicht haubare Holzbestände für die Waldwirtschaft besitzen' [On the Determination of the Value Which Forest Land and Immature Stands Possess for Forestry], *Allgemeine Forst- und Jagdzeitung,* pp. 441–55. [translated in *Oxford Institute Papers* No. 42 (1968)].

Helmedag, Fritz (2002) 'Die optimale Rotationsperiode erneuerbarer Ressourcen' [The Optimum Rotation Period of Renewable

Resources], in J. Backhaus and F. Helmedag (eds), *Holzwege: Forstpolitische Optionen auf dem Prüfstand*, Marburg, pp. 11–42.

Manz, Peter (1986) 'Forestry Economics in the Steady State: The Contribution of Johann Heinrich von Thünen', *History of Political Economy*, 18, pp. 281–90.

Nellinger, Ludwig (2002) 'Johann Heinrich von Thünens wirtschaftswissenschaftlicher Beitrag zu einer nachhaltigen Nutzung natürlicher Ressourcen' [Johann Heinrich von Thünen's Contribution to the Economics of the Sustainable Use of Natural Resources], in *Berichte über Landwirtschaft. Zeitschrift für Agrarpolitik und Landwirtschaft*, 215, pp. 39–56.

Samuelson, Paul A. (1976) 'Economics of Forestry in an Evolving Society', *Economic Inquiry*, 14(4), pp. 466–92.

Scorgie, Michael and Kennedy, John (1996) 'Who Discovered the Faustmann Condition?', *History of Political Economy* 28(1), pp. 77–80.

Van Suntum, Ulrich (1995) 'Johann Heinrich von Thünen als Kapitaltheoretiker' [Johann Heinrich von Thünen as a Theorist on Capital], in Heinz Rieter (ed.), *Studien zur Entwicklung der ökonomischen Theorie*, 14, Berlin: pp. 87–113.

Section One.

§ 1.
Timber yield.

After repeated observation, pines felled after 30 years at Tellow give on average 30 cartloads from 100 Mecklenburg square rods. I calculate the cartload to be on average, and including all branchwood, 64 Lübeck cubic feet.[1] As a rule, there are here:

7 cartloads of useable timber
23 cartloads of firewood

---

1. The Mecklenburg rod is equal to 16 Lübeck feet of 129 Parisian lines. The Magdeburg *Morgen* is equivalent to 117.9 Mecklenburg square rods. (100 Lübeck cubic feet = 79.1 Rhineland cubic feet and 100 Rhineland cubic feet = 126.4 Lübeck cubic feet). The timber yield of 100 Lübeck cubic feet over 100 Mecklenburg square rods is equal to the yield from 93.2 Rhineland cubic feet per Magdeburg *Morgen* and, in reverse, the yield of 100 Rhineland cubic feet per Magdeburg *Morgen* equals 107.2 Lübeck cubic feet over 100 Mecklenburg square rods.

Editorial addition: *1 Lübeck foot = 0.291024 m*
1 Mecklenburg rod = 4.6545 m
1 Parisian line = 2.256 mm
1 square rod = 21.66437 square metres
1 Waldmorgen = 2816.3681 square metres

1

The useable timber consists of slats for thatched roofs and fencing.

A cartload of firewood contains about a half cord of roundwood and a half cord of brushwood, reckoning the volume of the cord as 196 Lübeck cubic feet.

Based upon the views and investigations of Chief Forester Nagel of Diekhof, I assume in regard to the growth of wood the following:

With *appropriate thinning out*, through which the individual trees are *constantly* given space for the growth of the entire wood to achieve its maximum, from the sixth year the growth of timber in a pinewood remains constant and so the volume of timber increases at an arithmetic rate – excluding the oldest parts of the stand. Chief Forester Nagel and myself consider the first five years necessary for the establishment of young saplings and treat the timber of a six-year-old pinewood on local soil as equal to the annual timber growth during the years that immediately follow.

For Tellow we have calculated the 30-year-old pinewood at 30 cartloads per 64 cubic feet = 1,920 cubic feet on 100 square rods. This comes from the growth of 30 – 5 = 25 years. This gives an annual growth of 76.8 cubic feet on 100 square rods.

The area necessary for the realization of an annual growth of 100 cubic feet I call for the sake of brevity a *Waldmorgen*. The size of the *Waldmorgen* is here, where 100 square rods yields 76.8 cubic feet,

$$\frac{100}{76.8} \times 100 = 130 \text{ sq.rods}$$

§ 2.
Timber value.

A cartload of firewood contains:

a. a half cord of roundwood, which given the low local prices for wood[2] at 2 Thalers 4 shillings N⅔ = 1 Thaler 2 shillings N⅔.

b. a half cord of brushwood at

18 shillings = $\dfrac{9}{1\,Tlr\,11\,sh.}$

Felling costs, transport and bundling is estimated per cord at $\dfrac{21}{38\,sh.}$

The 23 cartloads of firewood thus have a value of 23×38 sh. = 874 sh.

The cartload of useable wood has at the price of 4 sh. for fencing and 2 sh. for thatching slats a value of 2½ Thalers or 120 sh.

For 7 cartloads this comes to	840 sh.
The value of 30 cartloads is therefore	1,714 sh.

This makes one cartload	
for 64 cubic feet on average	57 sh.
for 1 cubic foot on average	0.9 sh.
and for 100 cubic feet on average	90 sh.

Value of 100-year-old pine timber.

The price of strong building timber is hereabouts around 4 sh. per cubic foot.

Now if we assume that from a 100-year-old pine tree ⅔ of the mass is useable building wood and ⅓ firewood, we arrive at the following average value:

2 cubic feet at 4 sh.	= 8 sh.
1 cubic foot at 0.9 sh.	= 0.9 sh.
Which gives 3 cubic feet	= 8.9 sh.

2. When the Thaler is used here without qualification then Thaler N⅔ is meant, where 6 equal 7 Prussian Thaler Preußisch Courant. The Thaler N⅔ is equivalent to 48 shillings.

1 cubic foot is on average =   2.97 sh. N⅔.[3]

Increasing value of wood with age.

For a 100-year-old pine a cubic foot is   2.97 sh.
For a 30-year-old pine a cubic foot is   0.9 sh.
___

Difference of 70 years   2.07 sh.

This gives an annual increase in value of 0.03 sh. per cubic foot and 3 sh. N⅔ for 100 cubic feet.

Pine stocks of 30 years have per 100 cubic feet a value of 90 sh.

If we now apply the proposition derived from the difference in value between stands of 30 and 100 years – 'that in each year the value of 100 cubic feet increases by 3 sh. N⅔' – to younger stocks, we find that for 100 cubic feet the value

of 29-year-old wood is	87 sh.
of 28-year-old wood is	84 sh.
of 27-year-old wood is	81 sh.
...	...
of 6-year-old wood is	18 sh.

Remarkably, we can derive from the relationship in price between local young and old wood the following proposition:

That the value of 100 cubic feet of pine timber is related positively to the age of the wood.

___

3. Editorial addition: N⅔ = Neu ⅔ = New ⅔.

Hence both these elements are included in the increasing value, or the annual rent, of a wood:

1) the growth of timber,
2) the increased value of the existing timber stock.

§ 3.
The determination of the value of pinewood stock of a given age.

From the above we can calculate the value of stands of every age.

If the age of the stock = $x + 5$ years, the stock of timber = $100 \times$ cubic feet per *Waldmorgen*.

The value of 100 cubic feet is 3 sh. multiplied by the age of the tree, here therefore 3 $(x + 5)$ sh. The stock of $100 \times$ cubic feet has therefore a value of $3x (x + 5) = 3x^2 + 15x$, assuming the unit to be Shilling N$\frac{2}{3}$.

Example:

Age of felled wood	Wood stock in cubic feet	Value of 100 cubic feet (sh.)	Value of felled wood (sh.)	Sum of value of all felled wood (sh.)
6 years	100	18	18	...
7 years	200	21	42	60
8 years	300	24	72	132
9 years	400	27	108	240
10 years	500	30	150	390
11 years	600	33	198	588
12 years	700	36	252	840
13 years	800	39	312	1,152
14 years	900	42	378	1,530
15 years	1,000	45	450	1,980

From the continuation of this calculation the sum of values for all stocks in any rotation period can be established.

It remains to state a formula by means of which we can calculate this sum without working through each row of the table.

For this purpose let us consider more closely the ordering of the increase in the value of wood stock:

Age of wood stock	Value of wood stock	First difference	Second difference
6 years	18	...	...
7 years	42	24	...
8 years	72	30	6
9 years	108	36	6
10 years	150	42	6

For a series whose differences are ultimately constant this sum can be established by a general formula.

According to Euler (*Differentialrechnung* Part I Ch. 2 p. 56)[4] for a row whose first element $= a$
the first difference $= b$
the second difference $= c$
number of terms $= x$

$$\text{The sum } ax + \frac{x(x-1)}{1.2}b + \frac{x(x-1)(x-2)}{1.2.3}c$$

For the row which gives the value of stocks $a = 18$, $b = 24$, $c = 6$

---

4. Leonhard Euler, *Vollständige Anleitung zur Differentialrechnung*, Berlin 1793.

For these values of *a*, *b* and *c*

$$\frac{x(x-1)(x-2)}{1.2.3}6 = x^3 - 3x^2 + 2x$$

$$\frac{x(x-1)}{1.2}24 = 12x^2 - 12x$$

$$18x = \quad\quad + 18x$$

$$\text{Total} = x^3 + 9x^2 + 8x$$

Testing this:
If $x = 10$, then $x^3 = 1{,}000$
$$9x^2 = 900$$
$$8x = 80$$
$$\text{Summing} = 1{,}980,$$
which agrees with the above table.

§ 4.
The determination of the value of timber stocks of differing rotation periods.

A. Rotation of 21 years.

If $x + 5 = 21$, then $x = 16$
The value of all stocks is $x^3 + 9x^2 + 8x$
If $x = 16$, then
$$x^3 = 4{,}096$$
$$9x^2 = 2{,}304$$
$$8x = 128$$

Value of 6,528 sh. or 136 Thaler N⅔ for 21 *Waldmorgen* each of 130 square rods.
This gives per *Waldmorgen* 311 sh. = 6 Thaler 23 sh.
This gives per square rod 2.39 sh.

B. Rotation of 28 years.

If $x = 28 - 5 = 23$, then
$$x^3 = 12{,}167$$

$$9x^2 = \qquad 4{,}761$$
$$8x = \qquad \underline{184}$$

Value of timber stock of 17,112 sh. for 28 *Waldmorgen* each of 130 square rods.
This gives per *Waldmorgen* 611 sh. = 12 Thaler 35 sh.
This gives per square rods 4.70 sh.

C. Rotation of 35 years.

$x = 35 - 5 = 30$, then

$$x^3 = \qquad 27{,}000$$
$$9x^2 = \qquad 8{,}100$$
$$8x \ = \qquad \underline{240}$$

Value of timber stock of 35,340 sh. for 35 *Waldmorgen* each of 130 square rods.
This gives per *Waldmorgen* 1,010 sh. = 21 Thaler 2 sh.
This gives per square rod 7.77 sh.

D. Rotation of 42 years.

$x = 42 - 5 = 37$, then

$$x^3 \ = \qquad 50{,}653$$
$$9x^2 \ = \qquad 12{,}321$$
$$8x \ = \qquad \underline{296}$$

Value of timber stock of 63,270 sh. for 42 *Waldmorgen* each of 130 square rods.
This gives per *Waldmorgen* 1,506 sh. = 31 Thaler 18 sh.
This gives per square rod 11.58 sh.

E. Rotation of 49 years.

$x = 49 - 5 = 44$, then

$$x^3 \ = \qquad 85{,}184$$
$$9x^2 \ = \qquad 17{,}424$$
$$8x \ = \qquad \underline{352}$$

Value of timber stock of 102,960 sh. for 49 *Waldmorgen* each of 130 square rods.

This gives per *Waldmorgen* 2,101 sh. = 43 Thaler 37 sh.
This gives per square rod 16.16 sh.

Summary:

Rotation period in years	Value of timber per *Waldmorgen*	Value of timber per square rod
21	6 Tlr. 23 sh.	2.39 sh.
28	12 Tlr. 35 sh.	4.70 sh.
35	21 Tlr.　2 sh.	7.77 sh.
42	31 Tlr. 18 sh.	11.58 sh.
49	43 Tlr. 37 sh.	16.16 sh.

§ 5.
The rent of the woodland soil.

If for the time being we abstract from the revenues from thinning, related uses such as hunting and so forth, as well as the costs of inspection, then the land rent of woodland is equal to the value of clear felling, minus the interest on the value of all wood stocks and the costs of reseeding the cleared area.

The interest on the value of the timber stocks can be calculated by the above propositions.

Since seeding often fails completely, and even where it is successful requires partial reseeding, given local experience the costs of seeding cannot be set lower than 2 sh. per square rod, that is, 5 Thaler 20 sh. per *Waldmorgen* of 130 square rods.

With a rotation period of $x + 5$ years clear felling of $100 \times$ cubic feet of timber, the value of 100 cubic feet $= 3 (x + 5)$ shillings and the revenue from such felling is:

$$3x(x+5)\,\text{sh.} = 3x^2 + 15x.$$

Examples:

A. Rotation of 21 years.

For $x = 21 - 5 = 16$ the felled value $3x^2 = 768$
$$15x = 240$$

summing to 1,008 sh.

The value of the timber stock is from the above calculation 6,528 sh.

Interest at 4% on this =	261 sh.
Seeding costs per *Waldmorgen* at 2 sh. =	260 sh.
Subtraction	521 sh.
Gives a surplus of	487 sh.

21 *Waldmorgen* give 487 sh. land rent,
Which is per *Waldmorgen*   23.2 sh.
Which is per 100 square rods   17.8 sh.

B. Rotation of 28 years.

For $x = 28 - 5 =$ 23
$$3x^2 = 1,587$$
$$15x = 345$$
revenue   1,932 sh.

Value of the timber stock = 17,112 sh.

Interest at 4% on this	684 sh.
Seeding costs are	260 sh.
To be subtracted	944 sh.
Remains a rent of	988 sh.

Which is rent per *Waldmorgen* of   35.3 sh.
Which is per 100 square rods   27.1 sh.

C. Rotation of 35 years.

For $x = 35 - 5$ =   30
$$3x^2 = 2,700$$
$$15x = 450$$
Revenue   ——   3,150 sh.

The value of the timber stock is 35,340 sh.,

interest on this	1,414 sh.
The new seeding costs	260 sh.
To be subtracted	1,674 sh.

Gives a rent of 35 *Waldmorgen* of		1,476 sh.
Which is a rent per *Waldmorgen* of	42.2 sh.	
Which is a rent per 100 square feet of	32.5 sh.	

D. Rotation of 42 years.

Now $x = 42 - 5 = 37$

$$3x^2 = 4,107$$
$$15x = 555$$

Revenue	4,662 sh.
Value of the timber stock 63,270 sh.	
Interest at 4% on this	2,531 sh.
Seeding costs	260 sh.
To be subtracted	2,791 sh.
The rent of 42 *Waldmorgen* is	1,871 sh.
Which is rent per *Waldmorgen* of	44.5 sh.
Which is rent per 100 square rods	34.2 sh.

E. Rotation of 49 years.

$x = 49 - 5 = 44.$

$$3x^2 = 5,808$$
$$15x = 660$$

revenue	6,468 sh.
Value of the timber stock 102,960 sh.	
Interest at 4% on this is	4,118 sh.
Seeding costs	260 sh.
Subtraction	4,378 sh.
Rent of 49 *Waldmorgen*	2,090 sh.
Which is rent per *Waldmorgen* of	42.7 sh.
Which is rent per 100 square rods of	32.8 sh.

Summary:

Rotation period	Revenue (sh.)	Interest on timber value	Seeding costs (sh.)	Rent for entire area (sh.)	Rent per Waldmorgen (sh.)	Rent per 100 sq. rods. (sh.)
21 Years	1,008	261	260	487	23.2	17.8
28 Years	1,932	684	260	988	35.3	27.1
35 Years	3,150	1,414	260	1,476	42.2	32.5
42 Years	4,662	2,531	260	1,871	44.5	34.2
49 Years	6,468	4,118	260	2,090	42.7	32.8

§ 6.
Calculations related to the rotation period.

First Example: To present the rotation period for which the rent of the woodland area is at its maximum as a general formula.

The preceding examples show that the extension of the rotation period from 21 years to 28, 35 and 42 is related to an increase of the land rent, but that further extension to 49 years results in a diminution of the rent.

This demonstrates that there must be a point in time for which the rent is at its maximum.

This point can be discovered by a lengthy and wearisome process of trial and error; but in this process we do not discover the law itself. The presentation of a general formula expressing this law should therefore be of both scientific and practical interest.

The connections apparent in the preceding examples permit the presentation of such a general formula, as the following calculation shows.

Clear felling results in revenue of $3x^2 + 5x.$

From that must be deducted:

1) interest on the value of all timber stocks;
   the value of the timber stock is $x^3 + 9x^2 + 8x$
   which gives at 4% the interest $0.04x^3 + 0.36x^2 + 0.32x$
2) The cost of seeding a *Waldmorgen* of 130 sq. rods at 2 sh. = 260 sh.

Subtracting these expenses the rent of one *Waldmorgen* becomes

$$\frac{-0.04x^3 + 2.64x^2 + 14.68x - 260}{x + 5}$$

To find the value of x at which this function is at its maximum, we have of course to take its differential and set it equal to 0.
The differential is

$$(x + 5)(-0.12x^2 + 5.28x + 14.68)dx$$

$$-(-0.04x^3 + 2.64x^2 + 14.68x - 260)dx$$

Working through this formula gives:

$$-0.12x^3 + 5.28x^2 + 14.68x$$

$$-0.60x^2 + 26.40x + 73.40$$

$$\underline{+0.04x^3 - 2.64x^2 - 14.68x + 260}$$

$$(-0.08x^3 + 2.04x^2 + 26.40x + 333.40)\ dx = 0$$

This gives:

$$x^3 - 25.5x^2 - 330x - 4,167 = 0$$

To solve this equation the second term has to be eliminated, so for this purpose we set $x = y + 8.5$. This can then be rewritten as:

$$
\begin{aligned}
x^3 &= y^3 + 25.5y^2 + 216.75y + 614 \\
-25.5x^2 &= \quad\quad -25.5y^2 - 433.50y - 1{,}842 \\
-330\,x &= \quad\quad\quad\quad\quad\quad - 330.00y - 2{,}805 \\
-4{,}167 &= \quad\quad\quad\quad\quad\quad\quad\quad\quad - 4{,}167
\end{aligned}
$$

Summing to $y^3 - 546.75y - 8{,}200 = 0$

The Cardan Rule gives $y^3 = fy + g$

$$
y = \sqrt[3]{\left(\frac{g + \sqrt{\left(g^2 - \frac{4}{27}f^3\right)}}{2}\right)} + \sqrt[3]{\left(\frac{g - \sqrt{\left(g^2 - \frac{4}{27}f^3\right)}}{2}\right)}
$$

For our equation $y^3 = 546.75y + 8{,}200$ is

$f = 546.75$

$g = 8{,}200$

$f^3 = 163{,}443{,}326$

$\frac{4}{27}f^3 = 24{,}213{,}826$

$g^2 = 67{,}240{,}000$

$g^2 - \frac{4}{27}f^3 = 43{,}026{,}174$

$\sqrt{g^2 - \frac{4}{27}f^3} = 6{,}560$

$\dfrac{g + \sqrt{(g^2 - \frac{4}{27}f^3)}}{2} = \dfrac{8{,}200 + 6{,}560}{2} = 7{,}380$

$\dfrac{g + \sqrt{(g^2 - \frac{4}{27}f^3)}}{2} = \dfrac{8{,}200 - 6{,}560}{2} = 820$

$$\sqrt[3]{\left(\frac{g - \sqrt{\left(g^2 - \frac{4}{27}f^3\right)}}{2}\right)} = \sqrt[3]{7,380} = 19.47$$

$$\sqrt[3]{\left(\frac{g - \sqrt{\left(g^2 - \frac{4}{27}f^3\right)}}{2}\right)} = \sqrt[3]{820} = 9.36$$

This gives $y = 19.47 + 9.36 = 28.83$ and
$$x = y + 8.5 = 28.83 + 8.5 = 37.33$$

The age at clear felling equal to the rotation period is $x +$ 5 years.

This therefore gives the result that under the prices given for wood at different ages, *assuming that thinnings have no value*, the maximum for the land rent is at 37.33 + 5 = 42.33 years.

Second Example: To determine the age of stock at which the value of the growth just covers the interest on the value of this stock.

Given age of stock of $x + 5$ years, the value of the timber $= 3(x + 5)x = 3x^2 + 15x$

The value of timber one year younger is $3(x-1)^2 + 15(x-1)$

$$= 3x^2 - 6x + 3 + 15x - 15 = 3x^2 + 9x - 12.$$

One year's growth therefore contributes:

$$3x^2 + 15x - (3x^2 + 9x - 12) = 6x + 12$$

The interest on the value of felled timber of $x + 5$ years comes to

$$(3x^2 + 15x)\frac{4}{100} = 0.12x^2 + 0.6x.$$

Setting this interest equal to the value of the growth gives:

$$0.12x^2 + 0.6x = 6x + 12$$
$$0.12x^2 - 5.4x = 12$$
$$x^2 - 45x = 100$$
$$+(45/2)^2 = +506.25$$
$$x - 22.5 = \sqrt{606.25} = 24.62$$
$$x = 47.12$$

The age at felling where the interest on the value of the timber stock is equal to the value of the annual growth, and the land no longer bears any rent is $x + 5 = 52.12$ years.

Third Example: At what age of felled timber does timber growth, after subtracting the interest on the timber value in this felling, give the greatest surplus; or at what age of the tree does its felling give the highest rent?

As in the previous example the value of
the growth is $\qquad = 6x + 12$
The value of the stock at $(x + 5)$ years $\quad = 3x^2 + 15x$
Interest on this comes to $\qquad 0.12x^2 + 0.6x$
Subtracting this interest from the value of the stock at $(x + 5)$ years gives a rent of

$$= -0.12x^2 + 5.4x + 12.$$

Setting the differential of this function equal to 0 gives:

$$(-0.24x + 5.4)dx = 0$$

hence $0.24x = 5.4$
and $x = 22.5$

For $x = 22.5$, that is for the age of stock at 27½ years the rent for the individual stock has its maximum. But this does not provide a norm for the most advantageous rotation period; this can only be found from the sum of the rents of all stocks calculated in this way divided by the number of the stocks, which calculation gives the rotation period for which the entire wooded area gives the highest rent. Since younger stocks give a much lesser rent than at 27, the point in time at which the average of all stocks given the maximum rent must be more than that point in time at which an individual stock gives the highest rent.

Numerical Examples for Individual Years

If the stock is $x + 5$ years old, then	Value of the growth: $6x + 12$	Interest on the value of the timber stock: $0.12x^2 + 0.6x$	Surplus: $0.12x^2 +$ $5.4x + 12$
For			
$x = 1$	18	0.72	17.28
$x = 10$	72	18.00	54.00
$x = 21$	138	65.52	72.48
$x = 22.50$	147	74.25	72.75
$x = 24$	156	83.52	72.48
$x = 30$	192	126.00	66.00
$x = 42.33$	266	240.44	25.56
$x = 47$	294	293.28	0.72

To arrive at the rent we need to subtract from the surplus derived in this way the costs of seeding.

Doubt can arise respecting the correctness of a procedure by which the interest is calculated directly on timber stocks, hence from timber growth; and whether a more correct procedure would be to calculate the interest on timber value as of the preceding year, as in the procedure followed in regard to loaned capital, where interest is calculated in the succeeding year.

But if we rigorously separate the wages of labour from the utilization of capital we necessarily assume that wages are paid daily, since paying wages annually would include interest payments; so we here also assume that the interest on timber value is likewise paid daily, or calculated daily, since interest paid at the end of the year includes interest upon interest.

§ 7.
Thinning out.

Forestry science offers further interesting problems in relation to thinning out.

When thinning out one can give individual trees as much space as one wishes. The greater the space, the greater the growth of individual trees; but at the same time the number of trees growing on a given area of the wood diminishes. There must therefore be a point in the interval between trees at which the growth of the entire woodland is maximized.

Question 1.   Where is this point, and what measure should be used for the interval that the individual trees should be from each other if this maximum is to be achieved?

Question 2.   Do we here use the diameter or the wood mass of the tree as the measure of this interval?

Question 3.   How does the relative interval alter with the growth of individual trees and the growth of the entire wooded area?

As far as I am aware, these questions have barely been raised in forestry science, much less solved.

In the study that we have immediately available we are primarily interested in the question:

How do we determine the value of thinning out, and what is the most advantageous rotation period if we take account of thinnings?

Opinion among the most experienced foresters differs so greatly on the value of thinning in relation to the value of clear felling, or that of the remaining stock, that no reliance can be placed upon them whatsoever.

The yield and value of thinning depends mainly on the greater or lesser interval between trees at thinning, and it is probable that the different values given for the wood gained from thinning are due to differences among foresters in the principles applied to the choice of interval.

My own experiences are insufficient for the determination of any definite conclusions.

But it is less a matter of the accuracy of the numbers than of the representation of the method through which the question of the most advantageous rotation period can be determined. For if such a method can be found, for every case for which figures exist, varying from place to place as they will, the desired result can be deduced.

In this spirit I assume the following.

1) In pine stands no thinning is carried out in the first years after planting since in the first place no gain can be made from the growth in wood, and in the second the cost of thinning would exceed the value of the wood so gained.

2) The growth of pines from their 6th to their 10th year inclusive I assume to be $\frac{2}{3}$ of the growth in the following years. If for example the annual growth from the 11th year equals 150 cubic feet, so the growth from the 6th to the 10th year inclusive $= \frac{2}{3} \times 150 = 100$ cubic feet.

If each year 100 cubic feet of the total growth passes into the persisting stand, the five years noted above yield no wood from thinnings. If on the other hand only 75 cubic feet passes into the persisting stand, then an annual amount of 25 cubic feet can be gained by thinning from this five years' growth, together with 75 cubic feet from the growth in the years following.

3) If, as is here assumed, that from the 11th year the crowns of the pines mature, and the individual trunks take up the space at which the entire area provides a maximum of growth, then the growth of the 11-year-old trees will already limit the available space, and in the 12th year the greatest amount growth will not occur unless some of the trees are removed by thinning out. And this will happen in each of the following years.

Thinning in this way would of course be possible for pine stands that originated in natural seeding – where the trees grow at varying intervals from one another – rather than planting; but the costs of annual clearing would, especially with younger stands, exceed the value of the material gained, and so in practice one does not thin annually, but rather after periods of several years.

However, to present a general formula we require the assumption that individual trees always occupy the space for which the entire wood achieves its maximum growth in value. This presupposes that every year a part of the annual growth is taken away in thinning.

But since annually conducted thinning cannot be carried out in practice we have to ask ourselves whether the revenue from periodic thinning can be represented by a regular series linked to the value of the remaining timber.

The following serves to answer this: the present capital value of a future revenue can be exactly calculated if the rate of interest is given, and hence also the sum of capital values of all subsequent periodic revenues in one rotation period. By employing this form of calculation the capital value of the rent, which increases annually according to a specific rule, can be determined. The value of a capital can correspondingly also be resolved into an annually increasing rent. Hence the present capital value of all future thinnings can be represented with mathematical precision by an increasing series in which the terms are proportional to the value of remaining stock of timber.

Another objection that can be raised against the representation of the value of thinnings in a regularly increasing series is:

> That periodic thinning does not constantly give timber the normal amount of space required, but only in particular years; so that the annual growth in the value of timber cannot be a maximum, which is however the assumption of the desired formula.

This objection is entirely well founded, and the consequent deviation can be very significant when the intervals between thinnings are long. To set close limits to this deviation we must assume in the following calculation:

> That thinnings are carried out at close intervals, and that the timber is given the space that it would normally occupy in the middle of the period. On this assumption the variation between the practical outcome and our calculation will not be significant, and so I believe that for the time being such an abstraction can be made.

The next objective in this study is to investigate whether, and to what degree, the rotation period for which ground rent is at its maximum alters if the value of thinning is taken into account.

We can also establish this – for want of sufficient data – if we calculate with hypothetical figures in respect of the timber yield and the value of the wood from thinning.

These assumptions are employed in the following.

1) From the 11th year onward ⅓ of annual growth is removed in thinning, ⅔ of the timber growth remaining as stock. Since we have assumed that the annual increase of the stock is 100 cubic feet per *Waldmorgen*, the total growth for the year equals 150 cubic feet, the annual yield of thinning being therefore 50 cubic feet.

2) The value per cubic foot of the wood taken through thinning is less than that of the remaining stock of timber, for the following two reasons:

a. In woods that have formed from seedlings – which we here assume – thinning out removes trees already overshadowed together with those that will become overshadowed before the next thinning. This nearly always involves the weaker trees of a smaller diameter and which also have a lesser value per cubic foot.

b. The cutting, dragging out and removal of thinnings timber takes much more time and is more expensive than felling and removing whole areas. Since we are only considering the value that wood has after subtracting the costs of felling and preparation, thinnings timber must for this reason be given a lower value.

For both reasons I assume the value of thinnings timber to be ⅔ the value per cubic foot of the standing timber.

From this there follows:

From the 11th year onward the annual growth per *Waldmorgen* is 150 cubic feet; 50 cubic feet is removed from this by thinning.

This has a value of 50 × ⅔ = 33⅓ cubic feet of the remaining timber stock.

The annual yield by value of thinning is therefore $\frac{33\frac{1}{3}}{100}$ = ⅓ of the annual increase in value of the remaining timber.

Example:

According to the Table in § 3

Age of stock	Value of felling	Increase in value	Sum of increases in value
From $x + 5$ = 10 years	150	...	...
11 years	198	48	48
12 years	252	54	102
13 years	312	60	162
14 years	378	66	228
15 years	450	72	300

A quick glance at this table shows that the sum of the increases in value is always 150 less than the value of the timber stock itself. This is unavoidable; for the value of the timber stock consists of the sum of the increase in value of all stocks from the 6th to the $x + 5$th year, and since the growth from the 6th to the 11th year, which contributes 150, is here omitted, because during these years no thinning is carried out, if the stock is to grow each year by 100 cubic feet then the sum taken here into account must be less than the value of the timber stock by 150.

The value of the $x + 5'$ stock is according to § 4 = $3x^2 + 15x$. The increase in value in the stock from the 11th to the $(x + 5)'$ year is hence = $3x^2 + 15x - 150$.

The value of all thinnings is ⅓ of this sum, hence $x^2 + 5x - 50$.

Total revenue is therefore:

1) from clear felling $= 3x^2 + 15x$
2) from thinnings $\underline{ = x^2 + 5x - 50}$
   sum $= 4x^2 + 20x - 50$

From which has to be deducted

1) interest on the value of all timber stock $0.04x^3 + 0.36x^2 + 0.32x$
2) the costs of seeding one *Waldmorgen* of 130 sq. rods. at 2 sh. 260

The surplus is $= -0.04x^3 + 3.64x^2 + 19.68x - 310.$

Dividing this surplus by the number of stock gives a rent per *Waldmorgen* of

$$\frac{-0.04x^3 + 3.64x^2 + 19.68x - 310}{x + 5}$$

If $x = 30$, then

$-0.04x^3 =\ -4 \times 270\ =$		$-1{,}080$
$3.64x^2 =\ \ \ 9 \times 364\ =$	$+3{,}276$	
$19.68x =\ \ \ 3 \times 196.8 =$	590	
$-310\ \ \ \ \ =$		$-\ \ \ 310$
	$3{,}866$	$-1{,}390$ $= 2{,}476$

This, divided by the number of stocks $x + 5 = 35$ gives a rent per *Waldmorgen* of 130 sq. rods of 70.8 sh. per 100 square rods of 54.5 sh.

If $x = 40$, then

$-0.04x^3 =\ -4 \times 640\ =$		$-2{,}560$
$3.64x^2 =\ \ 16 \times 364\ =$	$+5{,}824$	
$19.68x =\ \ \ 4 \times 196.8 =$	787	

		$-$ 310	
-310 =			
	6,611	$-$ 2,870	= 3,741

Dividing by 45 gives a rent per *Waldmorgen* of 83.1 sh. per 100 square rods of 63.9 sh.

If x is 50 then

$-0.04x^3 = -4 \times 1{,}250 =$		$-5{,}000$	
$3.64x^2 = 25 \times 364$ =	$+9{,}100$		
$19.68x$ = 5×196.8 =	984		
-310 =		$-$ 310	
	10,084	$-5{,}310$	= 4,774

Dividing by 55 gives a rent per *Waldmorgen* of 86.8 sh. per 100 square rods of 66.8 sh.

If x is 60 then

$-0.04x^3 = -4 \times 2{,}160$ =		$-8{,}640$	
$3.64x^2$ = $36 \times$ 364 =	$+13{,}104$		
$19.68x$ = 6×196.8 =	1,181		
-310 =		-310	
	14,285	$-8{,}950$	= 5,335

Dividing by 65 gives a rent per *Waldmorgen* of 82.1 sh. per 100 square rods of 63.1 sh.

The rent for the rotation period is

	Per *Waldmorgen*	Per 100 sq. rods
For $x + 5 = 35$ years	70.8 sh.	54.5 sh.
45 years	83.1 sh.	63.9 sh.
55 years	86.8 sh.	66.8 sh.
65 years	82.1 sh.	63.1 sh.

We can therefore conclude, without the use of algebraic calculation as in § 6, that rent reaches its maximum with a rotation period of between 54 and 55 years.

But this is not yet the actual ground rent. For we have to add revenue from other uses such as hunting and pasture, and subtract from it the costs of administration and supervision that forestry requires if we are to arrive at the final ground rent. The costs of administration and supervision differ greatly according to the size and location of the forest, the latter increasing or decreasing the risk of theft of timber. After subtracting a small amount for additional use, I take them to be 8 sh. per *Waldmorgen*, or 6.2 sh. per 100 sq. rods.

§ 8.
The rent of woodland in comparison with ground rent.

Just as in agriculture we distinguish between estate rent and ground rent, the first involving the common product of capital and land, the latter being a portion of this estate rent remaining after the subtraction of interest on capital used in the construction of buildings; so likewise in forestry we can combine revenue from the land and the capital embodied in the timber stock as 'the rent of woodland', and distinguish this from ground rent.

Calculation of rent of woodland.

Revenue from felling contributes $3x^2 + 15x$
Revenue from thinnings $\underline{\quad x^2 + 5x - 50\quad}$
Revenue $4x^2 + 20x - 50$

Expenses come to:

Cost of seeding	260
Administration costs for $x + 5$ *Waldmorgen* at 8 sh.	$8x + 40$
Expenses	$8x + 300$
Having deducted these expenses the rent of the woodland is	$4x^2 + 12x - 350$

If $x = 30$ then

$4x^2 = 4 \times 900$	3,600
$12x = 12 \times 30$	360
-350	-350
Woodland rent for 35 *Morgen*	3,610

This amounts to 103.1 sh. per *Waldmorgen*

If $x = 40$ then

$4x^2 = 4 \times 1,600$	6,400
$12x = 12 \times 40$	480
-350	-350
Woodland rent for 45 *Morgen*	6,530

Divided by 45 this gives for 1 *Waldmorgen* 145.1 sh.

If $x = 50$ then

$4x^2 = 4 \times 2,500$	10,000
$12x = 12 \times 50$	600
-350	-350
Woodland rent for 55 *Morgen*	10,250

This gives for 1 *Waldmorgen* 186.4 sh.

If $x = 60$ then

$4x^2 = 4 \times 3,600$	14,400
$12x = 12 \times 60$	720
-350	-350
Woodland rent for 65 *Morgen*	14,770

This gives for 1 *Waldmorgen* 227.2 sh.

Comparison of woodland rent and ground rent.

Ground rent can be derived from § 5, if one subtracts from the rent given there 8 sh. administration costs per *Waldmorgen*.

Rotation period	Woodland rent for a *Waldmorgen* of	Ground rent for a *Waldmorgen* of	Relation between woodland and ground rent
35 years	103.1 sh.	62.8 sh.	100:61
45 years	145.1 sh.	75.1 sh.	100:52
55 years	186.4 sh.	78.8 sh.	100:42
65 years	227.2 sh.	74.1 sh.	100:33

It is shown here that an extension of the rotation period increases the woodland rent, while the ground rent is an ever-smaller portion of the woodland rent and reaches its maximum with a rotation of 55 years.

For the rotation period that gives the highest ground rent the *Waldmorgen* yields 78.8 sh. This equals 60.6 sh. per 100 sq. rods, and for the Mecklenburg measure of acreage, the *Last Acker* = 6,000 sq. rods (around 50 Magdeburg *Morgen*) 75.7 Thaler N$\frac{2}{3}$. It can be concluded that for a relation where the assumptions concerning timber yield and timber prices hold, land in agricultural use giving less than 56.5 Thaler N$\frac{2}{3}$ as land rent (not to be confused with lease) can find more profitable employment if planted with pines rather than continuing in use as agricultural land.

The question of where the frontier between land used for forestry and for grain lies is of practical importance for the larger landlord. The figures related to the calculation that decides this question vary quite markedly from location to location and can even diverge between two neighbouring estates. But the method according to which the calculation is done is applicable generally.

The natural frontier between wooded land and agricultural land is materially altered, however, if the chief aim of silviculture is the production of firewood, which is why we have also to consider this kind of relation.

§ 9.
Underwood or brushwood.

Meadows and low-lying fields unsuitable for afforestation provide some useable timber, but mostly firewood.

The rotation period here has, for two different reasons, to be much shorter than for upland areas; since

1) the value of firewood increases much less with its strength – that is, with age – than does forest timber, and
2) with increasing age the growth of new branches itself becomes of uneven quality and flawed.

For these reasons a 20-year rotation is usual here.

Having made my own observations in Tellow I have made the following calculations for the yield of light wood and timber:

A 20-year turnover on 100 sq. rods of light timber regularly yields on average at clearance

14 four-in-hand cartloads of firewood
And 1 four-in-hand cartload of useable wood.
This gives:

6½ cords of cracked and roundwood at 2 Th. 4 sh.	13 Th. 26 sh. N⅔
8 cords of brushwood at 20 sh.	3 Th. 16 sh.
½ cord of wood useable for shafts, ladders and fencing posts	3 Th.
For 15 cartloads the revenue will be	19 Th. 42 sh. N⅔

This will be per cartload 1 Th. 15.6 sh.

Felling, removal, transport and cording costs according to my calculations 21 sh. per cartload. Subtracting these costs the value of one cartload is 42.6 sh.
Treating the firewood separately, 14 cartloads gives an income of 16 Th. 42 sh.

This is per cartload	1 Th. 9.8 sh. N⅔
which after subtracting costs of	21 sh.

comes to 36.8 sh. for one cartload.
For 2,000 sq. rods of light wood, of which 100 sq. rods are annually cut, the surplus of revenue from firewood and useable wood amounts to 15 cartloads at 42.6 sh., which is 13 Th. 15 sh.
For 6,000 sq. rods this makes 38 Th. 45 sh.
To find the woodland rent we have to subtract from this revenue the following:

1) the costs of replanting for those plants that do not regrow, on 6,000 sq. rods 3 Th.
2) cutting back hops in the younger plantings to protect their growth – around 2 Th.
3) Costs of administration and supervision per 100 sq. rods 6.2 sh. is for 6,000 sq. rods 7 Th. 36 sh.

Total expenses come to 12 Th. 36 sh. Subtracting these for 6,000 sq. rods gives a woodland rent of 27 Th. 9 sh.
Ground rent can be determined if the interest on the capital embodied by the standing timber is deducted from the woodland rent.
For 100 sq. rods the wood resulting from clear felling 100 sq. rods = 15 cartloads. If, as I believe can be assumed without serious error, the annual growth of timber from the first to the 20th year is constant, then the average timber yielded from all felling and cutting is 7½ cartloads. Therefore the entire timber stock of all 20 wooded areas on 2,000 sq. rods amounts to 150 cartloads. The

value of cartload from clear felling = 42.6 sh. The average value of the timber from all stands may be roughly
⅔ this sum, hence 28.4 sh. per cartload.

Accordingly, the value of all timber from 20 fellings on
100 sq. rods = 2,000 sq. rods is 150 × 28.4 sh. – 88 Th.
36 sh.

Hence for an area of 6,000 sq. rods
the capital value of the timber is   266 Th. 12 sh.
   With 4% interest this is   10 Th. 31 sh.,
   and the woodland rent is   27 Th.   9 sh.

Subtract the interest, for 6,000 sq. rods of light wood
we get a ground rent of 16 Th. 26 sh.

This rent from 6,000 sq. rods of light wood is so small
that one is tempted to assume that on an estate all such
light wood should be cleared from land that has some
value as arable, meadow or pasture.

However, from the moment that the timber needs of
the estate cannot be met from its own land, but has to be
purchased, there is a remarkable change in the cost of
timber for the estate.

If timber has to be bought in, then it is not only the
timber itself which is paid for, but the cost of transport.
Together with this there arises great uncertainty in
meeting the annually recurring need for timber. One
becomes dependent on the mood of one's neighbour,
and if he does not wish to sell wood, or if he decides to
clear out all his surplus timber, one can be in the position of having to purchase and transport timber from
miles away.

In this way the costs of purchase and transport can
raise the price of timber for the estate by a factor of two
or even three times the previous selling price.

In this regard it can be advisable – if the need for firewood cannot be met by collecting fallen wood – to culti-
vate light wood on land which, used as meadow or pasture,
gives from 6,000 sq. rods a rent of 40 Th. and above.

It is in any case advisable under these conditions to produce only as much firewood as is needed by the estate.

Since all estate owners have a common interest in clearing areas that produce firewood surplus to requirements, a shortage of firewood will consequently arise, and the contrast between the production price – including the rent that the woodland could give if used for arable or meadow – and the sale price will exist for just so long as there are areas of ancient woodland in existence that have been created by nature, and not the hand of man.

§ 10.
The influence of the yield in value of thinnings on the most advantageous rotation period and on ground rent.

Our study is primarily directed at investigation of the rotation period for which, *at a given timber price*, forestry yields the highest ground rent.

If this task is solved, the study can open the way for a general and higher task: to determine the rotation period for which the production price of timber is minimized.

At the outset of this study we assumed the regular occurrence of thinning, so that we could shed light on the question of whether, and to what extent, the value of thinnings affected the most advantageous rotation period. This is because without thinning the annual growth in value cannot be maximized. But we have abstracted from the yield the value of thinning; or in other words, we have assumed that the revenue from thinning is absorbed by the costs of felling and processing.

Further study was then based on the assumption that thinnings produce a net revenue and have assumed figures for this net revenue that may come close to reality in many cases.

Comparison of the results of these calculations contributes to answering the above question.

a. If thinnings yield no net revenue
In this case the most advantageous rotation period is, according to in § 5, 42 years.

Ground rent is then also as in § 5, 44.5 sh. per *Waldmorgen* minus 8 sh. for administration costs = 36.5 sh.

b. If thinnings have the value given above
Then the most advantageous rotation period = 55 years and the ground rent is then 78.8 sh. per *Waldmorgen*.

From this comparison we can see that by adding the net revenue of thinning:

1) the rotation period lengthens from 42 to 55 years, and
2) ground rent increases from 36.5 sh. per *Waldmorgen* to 78.8 sh.

It is however quite striking and interesting that even given the marked increase in the value of timber per cubic foot related directly to the age of the tree, a rotation period longer than 55 years proves disadvantageous.

But if all owners of woodland in an entire region, recognizing their own interest, do not permit lengthy rotations, the production of strong building timber is entirely neglected. And since such strong building timber is indispensable the production of it has to be stimulated and rendered advantageous by an increase in price.

Hence it appears that the timber price per cubic foot must increase much more markedly with the age of the tree than we have assumed if the costs of production of strong building timber is to be met.

We are therefore brought through our investigation to the understanding that if the value of thinning increases from zero to $\frac{1}{3}$ of the increase in value of the remaining timber stock, then the rotation period, in the absence of an increase in the price of older timber, is extended from 42 to 55 years.

Now the value of thinning is in no respect a constant in relation to the value of the remaining timber stock, but is entirely dependent on the method employed in thinning.

The question therefore automatically arises of whether there is not a method for thinning for which the increase in value of the thinning in relation to the value of the remaining stock extends the most advantageous rotation period such that strong building timber can be grown with advantage, without there being an increase in the relative price of young and older wood.

§ 11.
The thinning methods employed by Chief Forester Nagel.

In 1825 Chief Forester Nagel – a clear-thinking, practical forester[5] – published in the *Mecklenburgische Annalen der Landwirtschaft* Jg. 12, Part 2 an estimable calculation of woodland value that is of great value to me, since it gave me my first insights into the revenue of woodland and the methods of thinning.

I gladly present here what I gained from this essay and from the verbal clarifications and communications concerning thinning methods that Chief Forester Nagel gave me, since, given the advanced age of this forester, the results of his extensive studies would otherwise very probably have been lost to the public.

---

5. Who has since died.

This thinning method is based on the following propositions and positions:

1) For the purposes of the calculation it is assumed that a pinewood has not been naturally seeded, but planted in straight rows.
2) In the first thinning out, done when the pines reach the age of 15, every alternate row is removed. At the second thinning out, at 21 years, every other tree in each row is removed. This procedure is followed in the subsequent thinnings.
3) When the trees standing on a square have grown to the extent that the interval between trees is only ten times their diameter they are thinned in such a way that half of the trunks are cut. This procedure is followed with the:
    3rd thinning at age 30
    4th thinning at age 42
    5th thinning at age 60
    6th thinning at age 84
    And felled at 120 years
4) The growth in timber is, under this regime of thinning, equally great among younger and older stocks, and the timber stock grows at an arithmetical rate. A number of years is presumed to be needed for the establishment of the tree and for these years no growth is assumed; this varies according to the qualities of the land.

We now turn to the question:

What influence does this thinning method have on ground rent and the most advantageous rotation period?

For the time being we leave aside the question of whether this type of thinning promotes the greatest annual growth, and use the earlier figure for annual growth of 150 cubic feet per *Waldmorgen.*

In this case, however, half of the annual growth is removed by thinning, rather than $\frac{1}{3}$ as previously, hence 75 and not 50 cubic feet, and the remaining stock of timber increases not by 100 cubic feet per year, but 75 cubic feet.

Since here it is not the weaker trees subjected to being overshadowed that determines the selection of trees for removal but the space that they occupy, the timber yielded from thinning out must have a price equal to that of the timber stocks left standing.

Since, moreover, as a result of this method of thinning there very quickly emerges a network of accessible tracks, the removal of timber is rendered much easier and cheaper; and so we can assume, without great deviation from reality, that the thinnings timber has not only the same selling price but the same value per cubic foot as the remaining timber stock.

Taking these definitions we can derive the following general formula for the rent:

Given the age at felling of $x + 5$ years the stock of timber is 75 × cubic feet.
100 cubic feet have a value of $3x + 15$
This gives for 75 × cubic feet $2.25x^2 + 11.25x$

Timber from thinning therefore has the same value as felled wood, with the difference that those trees cut from the 6th to the 10th year inclusive, which are assumed in § 7 to have no thinnings value. Therefore the thinnings value has to be deducted from the value of the felled timber.

This deduction is $150 \times \dfrac{75}{100} = 112.5$

Accordingly, the value of all thinning is $2.25x^2 + 11.25x - 112.5$[6]

Felling and thinning together give $4.50x^2 + 22.50x - 112.5$

From this has to be deducted

1) The interest on the value of all standing timber. Given an annual growth of 100 cubic feet per year the interest amounts to:

$$0.04x^3 + 0.36x^2 + 0.32x$$

With an annual growth of standing timber of only 75 cubic feet the interest is only ¾ so high and hence is:

$$0.03x^3 + 0.27x^2 + 0.24x$$

2) The planting costs, which we for the time being set equal to the seeding costs, are 260. Deducting from this the rent for $x = 5$ *Waldmorgen* equals

$$-0.03x^3 + 4.23x^2 + 22.26x - 372.5$$

If one now replaces x with the sequence 50, 60, 70, 80 and 90, and deducts 8 sh. per *Waldmorgen* for administration costs, the calculated ground rent per *Waldmorgen* is

Rotation of 55 years	129.6 sh.
Rotation of 65 years	141.4 sh.
Rotation of 75 years	147.0 sh.
Rotation of 85 years	143.8 sh.
Rotation of 95 years	139.5 sh.

---

6. Here there is an error that I only later saw and hence passes into the following paragraphs; the more exact calculation, as carried out in § 21, gives the revenue from thinning with the formula $2.25x^2 + 11.25x - 75$. The variation that results here is however insignificant, and never in the following amounts to 1 sh. per *Waldmorgen*.

The maximum of the ground rent will fall between the 75th and the 76th year.

For the thinning method initially assumed we found (in § 7) that a rotation of 55 years gave the highest ground rent per *Waldmorgen*, at 78.8.

Here we find that this figure is now with a 55 year rotation 129.6 and for a 75 year rotation 147.

Nagel's method of thinning therefore seems to lead to an enormous increase in the ground rent.

But this comparison cannot provide a norm for our investigation, for the following reasons:

1) If thinning is to be profitable after just 15 years, the planting has to be extraordinarily dense, and so the costs of planting will quite considerably exceed the costs of seeding – the only costs here considered. If the planting is less dense, however, the first years will fall far short of the anticipated growth and the first thinning can then be best done in the 30th year.

2) Since with this method the incidence of thinnings in the mature years is progressively less, then shortly before the thinning the trees will be too crowded, and afterwards, when each tree has gained double the space, there will be too much space for each tree. Therefore, the maximum of growth assumed in this calculation cannot be achieved.

3) If trees are during thinning only assessed with regard to the space that they occupy, then many bent or badly grown trees will be left standing, so that when the trees are finally felled there will be more firewood and less building timber than we have assumed in our calculations.

I would not wish to estimate by how much the calculated ground rent would be diminished by these circumstances. But it is indisputable that the fact that capital embodied in the standing timber, and the associated interest, are both ¼ less than with the first thinning method (§ 7), is a very important reason for the higher ground rent that Nagel's method gives.

To properly determine the influence of the reduction in the timber capital on ground rent, without the influence of disturbing incidental circumstances, we have to apply the proposition of removing half the growth through thinning to seedlings.

§ 12.
Ground rent and rotation period from pollination, if thinning removes half the wooded growth.

While seeding does not permit seedlings to assume an entirely uniform, rectangular spacing of the trees, there is no doubt that more vigorous clearing during thinning can clear away one half of the growth, as in the case of plantings.

The problem with Nagel's method, that trunks are either too close to each other, or have too much space, can be almost entirely removed here if shorter thinning periods are adopted. In the following I therefore assume that the first thinning occurs in the 15th year, the second in the 25th year, the third in the 35th year and so on every ten years. By contrast, the advantage of Nagel's method that the thinned wood has the same value per cubic feet and the remaining standing timber cannot be achieved with self-seeding. The reason, as already suggested, is that thinning will be directed to the less valuable growth, and at the same time the retrieval of the wood becomes that much more expensive, because of the cost of cutting and dragging it out of the wood. As before, we therefore set the value of thinning wood per cubic foot at $\frac{2}{3}$ the value of the value of the remaining timber.

Accordingly, the value of thinning comes to

$$\frac{2}{3} \ (2.25x^2 + 11.25x - 112\frac{1}{2})$$

Hence $0.75x^2 + 3.75x - 37\frac{1}{2}$ less than with Nagel's method.

Revenue from felled timber, together with the expenses for interest and seeding remain unchanged.

In the preceding section the rent was

$$- 0.03x^3 + 4.23x^2 + 22.26x - 372\frac{1}{2}$$

The reduction in
revenue from thinning is $\underline{\hspace{2em} 0.75x^2 + 3.75x - 37\frac{1}{2}}$

The rent for pollinated
land is therefore $\quad -0.03x^3 + 3.48x^2 + 18.51x - 335$

If, as in the preceding section, one sets x equal to a series of values, and deducts 8 sh. administration costs per *Waldmorgen*, we have the following results: Ground rent is per *Waldmorgen*

Rotation of 55 years	92.7 sh.
Rotation of 65 years	97.0 sh.
Rotation of 67 years	97.1 sh.
Rotation of 75 years	95.0 sh.

The highest ground rent occurs at a rotation period of 67 years.

Woodland rent is

$$3.75x^2 + 18.75x - 335$$

Deducting 8 sh. for administration costs, this gives the following:

Rotation of 55 years	173.4 sh.
Rotation of 65 years	211.8 sh.
Rotation of 75 years	250.0 sh.

Relation of woodland rent and ground rent.

Rotation period	Woodland rent	Ground rent	Relation
55 Years	173.4	92.7	100:53.5
65 Years	211.8	97	100:45.8
75 Years	250.0	95	100:38.0

§ 13.
Comparison of the two methods, in which thinning removes

A. One-third of annual timber growth,

B. Half of this growth

1. Concerning the part of the volume, which is captured by thinning.

If the first thinning takes place when the trees are 15 years old, and this is then repeated every ten years, the thinned wood relative to existing stock is:

In case A, if $\frac{1}{3}$ of the growth is removed through thinning.

At age 15 the stand is $5 \times 100 + 5 \times 150 = 500 + 750 = 1,250$ cubic feet then the annual growth from the 5th to the 10th year = 100 cubic feet and from the 10th to the 15th year = 150 cubic feet.

Thinning removes from this $5 \times 50 = 250$ cubic feet; hence $\frac{1}{5}$ of the stock.

The remaining stock	= 1,000 cubic feet
To which is added the growth in ten years	= 1,500 cubic feet
Hence at age 25	= 2,500 cubic feet

From this the second thinning removes $10 \times 50 = 500$ cubic feet, that is, one-fifth of the stock.

The remaining stock = 2,000 cubic feet
Added to this ten years' growth = 1,500 cubic feet

Stock at age 35 = 3,500 cubic feet

Thinning again removes 500 cubic feet, hence $\frac{1}{7}$ of the total stock.

Projecting this calculation, thinning removes:

At age 45	$\frac{1}{9}$
At age 55	$\frac{1}{11}$
At age 65	$\frac{1}{13}$
At age 75	$\frac{1}{15}$

In case B, if half of the growth is removed through thinning.

At age 15 the stand = 1,250 cubic feet.

The yield of thinning is $5 \times 25 + 5 \times 75 = 125 + 375 = 500$, hence $\frac{2}{5}$ of the stock.

For the stand there remains 750 cubic feet
Growth in 10 years 1,500 cubic feet

Stock at age 25 2,500 cubic feet

Thinning takes away $10 \times 75 = 750$, hence $\frac{1}{3}$ of the stock.

This leads to the following series:

The 3rd thinning at age 35 takes $\frac{1}{4}$ of the stock
The 4th thinning at age 45 takes $\frac{1}{5}$ of the stock
The 5th thinning at age 55 takes $\frac{1}{6}$ of the stock
The 6th thinning at age 65 takes $\frac{1}{7}$ of the stock
The 7th thinning at age 75 takes $\frac{1}{8}$ of the stock.

2. Concerning the rent of woodland.

Woodland rent is per *Waldmorgen*:

	For method A	For method B
35 year rotation period	103.1	...
45 year rotation period	145.1	...
55 year rotation period	186.4	173.4
65 year rotation period	227.2	211.8
75 year rotation period	...	250.0
...	...	...
105 year rotation period	389.0	363.8

From this it appears that:

1) Method A produces a greater woodland rent than Method B.
2) The woodland rent constantly and substantially increases with the longer rotation period.

It becomes plain from this result, which can also be established by a simpler calculation, why less intensive thinning and a longer rotation period has been adhered to for so long and so stubbornly.

But in fact this calculation shows just one thing: that the interest on a large amount of capital is more than that on a smaller!

For with the longer rotation by far the greatest part of the woodland rent derives from the interest upon the capital represented by the timber stock.

For A, interest on the value of all timber stocks contributes (§ 7) $0.04x^3 + 0.36x^2 + 0.32x$.

For $x = 100$ and a rotation of 105 years this gives interest of 415.5 per *Waldmorgen*.

Interest on the capital embodied by the timber stock therefore here exceeds the entire woodland rent, however considerable this last might be – and the owner of the woodland would, if he felled all the timber, invested the capital realized for interest and left the land entirely untended, have a greater revenue than he hitherto had from silviculture.

3. Concerning the ground rent.

Ground rent is per *Waldmorgen*:

	For method A	For method B
35 year rotation period	62.8	...
45 year rotation period	75.1	...
55 year rotation period	78.8	92.7
65 year rotation period	74.1	97.0
75 year rotation period	...	97.1
...	...	95.0
105 year rotation period	−26.5	52.2

From this it is apparent that the highest ground rent that Method A can yield is 78.8 sh. per *Waldmorgen* of 130 square rods, which for an area of 6,000 square rods would be 75.7 Th.; the highest ground rent that Method can yield is by contrast 97.1 sh. per *Waldmorgen* = 93.3 Th. for 6,000 square rods. The proportion between A and B is 78.8:97.1 = 100:123; or B yields 23% more ground rent than A.

There is an additional circumstance that benefits Method B and which has not here been taken into account. Where trees have greater clear space the growth of the individual trunks is greater even if the total growth of the woodland remains the same. The increasing value per cubic foot of the timber with age is, however, only a result of the increasing diameter of the older trunks, and if for different thinning methods the diameter of the trees varies, then the diameter of the tree should become the standard for the value per cubic foot of the timber, and not the tree's age.

Assuming that with Method A the diameter of the individual trees increases annually by $\frac{1}{6}$ of an inch, with Method B by $\frac{1}{5}$ of an inch, with A the tree will arrive at a diameter of 1 foot at 72 years, while with Method B this will be reached as soon as 60 years; and both trees will have the same value, disregarding the difference in ages.

Assuming that for A an annual increase in value of timber of 0.03 sh. per cubic foot is appropriate, then for B the increase in value can be estimated as 0.036 sh. per cubic foot.

If we take the figures from these worked examples as a norm, then the surplus of the yield of B with respect to A climbs to almost 50%.

However, if a substantiated calculation is to be made, one needs to know the law according to which the diameter of the trunks increases given different amounts of clear space.

Nonetheless, the result established here deserves consideration. For if the net yield of pinewood can be increased by an amount between 23% to 50% through an improved thinning method, this is not only of use to the owner of the woodland, but national income will itself also increase by a significant amount.

Nevertheless, this investigation is incomplete.

For B as well as for A the annual growth per *Waldmorgen* is assumed to be 150 cubic feet. There is, however, no doubt that the greater or smaller amount of clear space between trees not only has an effect on the increase in size of individual trees but also has an influence on the total growth of the entire wood; and so we cannot assume one and the same annual growth for two thinning methods.

The question, however, of whether the annual growth is greater with A or with B, and in what proportion these two methods stand, is insoluble as long as we do not know the relation of the growth of individual trees to the space which each tree is given.

This demonstrates again that the progress of knowledge leads not to perfect results but rather to problems of a higher level.

4. Concerning the rotation period.

We have seen from the preceding investigations that the most advantageous rotation period alters with the value and yield of thinning, and that the land yields the highest rent:

a) when the wood yielded from thinning has no value: a rotation period of 42 years;
b) if thinning takes $\frac{1}{3}$ of the timber growth and the wood so yielded has $\frac{2}{3}$ the value of the remaining timber: a rotation period of 55 years;
c) if thinning takes half of the timber growth, and the wood so yielded has $\frac{2}{3}$ the value of the remaining timber: a rotation period of 62 years.

It quite definitely emerges from this that the rotation period that produces the greatest profit is the longer, the greater the amount yielded from the thinning.

It is nonetheless striking that even vigorous thinning taking one half of the timber growth does not extend the most advantageous rotation period beyond 67 years. Pines of 67 years still do not yield strong building timber, which appears therefore to confirm the view expressed earlier that the price of building timber per cubic foot has to increase proportionally more than the age of the tree if strong building timber is still to be produced after the disappearance of ancient woodland.

If we now consider more closely what prevents a longer rotation period and renders it disadvantageous, we find the reason in the large amount of capital embodied in the timber stock with a lengthy rotation period. This is especially apparent with a 105 year rotation, where with Method A the interest from the stock does not only eat up the entire woodland rent, but exceeds it by 26.5 sh. per *Waldmorgen*.

By contrast, the same rotation with Method B, where the capital stock is ¼ less, still gives a ground rent of 52.2 sh. This teaches us that the disadvantages of higher rotation periods can be diminished by the reduction of the stock, that is, by giving each tree more space when thinning.

How far such a practice might be taken is the object of the following study.

Section Two.

§ 14.
How great does the space around each tree need to be in relation to its diameter if the annual growth in value of the entire woodland is to achieve a maximum?

This relation will vary according to the different kinds of tree, and will vary even for the same type of tree according to the ground on which it stands. Thus it seems that in this respect no rule can be devised.

But we are not talking here about one rule that covers all types of tree and land; we instead consider one type of tree (the pine) on a particular type of land, seeking to discover a law and from that develop further conclusions.

If one wishes to claim that there is not even a law that covers this specific case, then my response is: will nature fail to provide an answer if an experiment is constructed as in the following?

A field of perfectly uniform soil is divided into eight equal parts, is sown uniformly with pines, and so far as is possible in the annual thinning, the trunks are spaced, from their youth to the time they are felled, according to the following sequence:

In the 1st section	8 times their diameter
In the 2nd section	9 times their diameter
In the 3rd section	10 times their diameter
In the 4th section	11 times their diameter
In the 5th section	12 times their diameter
In the 6th section	13 times their diameter
In the 7th section	14 times their diameter
In the 8th section	15 times their diameter

Nature will certainly give an answer here; and if we repeat this experiment on the same land under the same conditions, so that we might average out possible disturbances, then for this concrete case a law will be established.

Since there is no absolute obstacle to the setting up of such an experiment for any type of tree on any kind of land and under all climatic conditions, the possibility of investigating this natural law cannot be denied, apart from the great difficulty of such an experiment.

Setting up experiments of this kind demands a great deal of attention on the part of the observer, a significant cost and, what is worse, a period of time that exceeds the lifetime of an individual person. It is no wonder that a resolution to this question remains unaccomplished. But I have been struck by the fact that nowhere in the forestry literature with which I am familiar have I found any figures concerning the space required for a given diameter of a tree's trunk, not even hypothetical figures – as any principle governing thinning out must ultimately rest on this question.

The only useful elements that I have found on this subject are in the thinning methods of Chief Forester Nagel.

As already outlined in § 11, he carried out thinning when the interval between trees was ten times their diameter, and then by removing half of the trees gave the remaining trunks twice the amount of space.

If we here take the square of the diameter = δ^2 as the unit of measurement for this space, then each tree, shortly before thinning, has 100 δ^2, and afterwards 200 such units.

The relative space of 200 δ^2 for each tree with which each thinning period begins does however diminish each year, since the unit with which we are measuring – the diameter – grows constantly, until at the end of this thinning period the relative space has sunk to 100 δ^2.

But if we follow this procedure the trees have too much space at the beginning and too little at the end of the period, and from this we can only deduce that the trees are at a normal distance in the middle of this period.

In the middle of the period the tree has a space of

$$\frac{200\,\delta^2 + 100\,\delta^2}{2} = 150\,\delta^2$$

When the space available is 150 δ^2 the interval between trees is

$$= \sqrt{150\,\delta^2} = 12.15\,\delta$$

Accordingly, further discussion is based upon the following assumption:

If an area of woodland is to yield the highest annual increase in value, the interval between each tree from the mid-point of the area occupied by one tree to the mid-point of those trees in the surrounding square must be twelve times its diameter.

I have to leave verification and probable correction of this assumption to practical foresters.

§ 15.
The growth of individual trees in diameter and in physical volume.

Chief Forester Nagel assumes that if the interval separating one tree from the next is normal, then the annual growth in their diameter is arithmetical and is equal in every year, leaving aside the early years when the tree is establishing itself and excluding those when it is very old.

Example: The number of years needed for the tree's establishment and for which no growth is imputed in our calculation is five, and the annual growth in the diameter is, from the 6th year on, $\frac{1}{5}$ of an inch or $\frac{1}{60}$ of a foot; and so the diameter of the tree is

$$
\begin{array}{llll}
\text{Aged 25} & 20 \times \tfrac{1}{5} = & 4 \text{ inches} \\
\text{Aged 35} & 30 \times \tfrac{1}{5} = & 6 \text{ inches} \\
\text{Aged 45} & 40 \times \tfrac{1}{5} = & 8 \text{ inches} \\
\text{Aged 55} & 50 \times \tfrac{1}{5} = & 10 \text{ inches} \\
\text{Aged 65} & 60 \times \tfrac{1}{5} = & 12 \text{ inches}
\end{array}
$$

Calculation of the physical volume of a tree 1 foot in diameter and 71 feet high.

The tree is a truncated cone with branches, but which is in the middle usually rather thicker than required by the pure conic form.

Cotta has devised tables on the relation between the volume of a tree and that of a mathematical cone occupying the same height and base area, greatly facilitating our calculations.[1]

---

1. Editorial addition: Heinrich Cotta, *Hiilfstafeln für Forstwirthe und Forttaxatoren*, Dresden 1821.

According to Table I, p. 5 and Table II, p. 23, the volume of a cone with a diameter of 1 foot and a height of 71 feet = 18.59 cubic feet.

According to Table IV, p. 32, for pines the average proportion between the volume of the cone and the volume of a tree stripped of its branches is 100:129. Hence the volume of a 71 foot high pine of 1 foot diameter is, without branches, $18.59 \times 1.29 = 23.98$ cubic feet.

The calculation is much simplified, however, if we take a four-sided pyramid rather than a cone as the standard of comparison.

If the relation of diameter to height is 1:h, then for a diameter of δ the height is $h\delta$. The pyramid, whose basal area is a square, each side of which is equal to the diameter of the tree, hence δ^2, has a volume of $\frac{1}{3}h\delta \times \delta^2 = \frac{1}{3}h\delta^3$

For $h = 71$ and $\delta = 1$ the volume of the pyramid is $\frac{1}{3} \times 71 = 23.67$ cubic feet. The volume of the tree is = 23.98 cubic feet.

There is a difference here of 0.31 cubic feet, and to find the volume of a tree stripped of branches the volume of the pyramid has to be multiplied by 1.013.

With full stems, that is, boles with all branches, the proportion of the volume of the whole tree to that of the cone is (according to Cotta's Table 4, p. 32) = 167:100.

The volume of the cone is 18.59 cubic feet. The volume of the entire tree is therefore $18.59 \times 1.67 = 31.05$ cubic feet.

If we compare this to the volume of a pyramid, the proportion between them is 23.67:31.05 = 100:131.2.

Hence the volume of the tree with all its branches can be established if one multiplies the corresponding pyramid by 1.312.

The volume of the pyramid is $\frac{1}{3}h\delta \times \delta^2 = \frac{1}{3}h\delta^3$

The volume of the tree is then $\dfrac{1.31}{3}h\delta^3$

For $h = 71$, $\dfrac{1.31 \times h}{3} = \dfrac{1.31 \times 71}{3} = 31$ and we establish that the volume of the tree can be expressed by the very simple expression $31\delta^3$

This gives for $\delta = 2$ the volume $= 248$ cubic feet
This gives for $\delta = 1$ the volume $= 31$ cubic feet
This gives for $\delta = \frac{1}{2}$ the volume $= \frac{27}{8}$ cubic feet

If the annual increase in diameter of the tree is $\frac{1}{6}$ of an inch $= \frac{1}{72}$ feet, how great is the increase in physical volume of a tree with a 1 foot diameter?

The tree that in spring is 1 foot broad will in the autumn of the same year have a diameter of $1\frac{1}{72}$ feet $= \frac{73}{72}$ feet.

The volume of this tree is

$$31\left(\frac{73}{72}\right)^3 = 31 \times \frac{389,017}{373,248} = 32.31\,\text{cu.ft.}$$

The volume of the tree has increased from 31 cubic feet to 32.31 cubic feet, the growth being 1.31 cubic feet or $\dfrac{1.31}{31} = 4.2\%$ of the stock.

In general, if the annual growth of the diameter is α feet, in the course of the summer the diameter increases from δ to $\delta + \alpha$.

The volume of a tree whose diameter $= \delta + \alpha$ is $31(\delta + \alpha)^3 = 31(\delta^3 + 3\delta^2\alpha + 3\alpha^2\delta + \alpha^3)$.

Early in the year its volume was $31\delta^3$.

The increase of physical volume is therefore $31(3\delta^2\alpha + 3\delta\alpha^2 + \alpha^3)$.

§ 16.
Calculation of the portion of the growth to be removed in thinning.

Given the great difference of opinion among foresters on the yield of thinning, and the complete lack of tried and tested propositions for thinning, I have in the fore-going only been able to assume hypothetically the yield of thinning in comparison with growth.

Given this deficiency in our knowledge of one of the most important objects of forestry economics, it is as little possible to determine the most advantageous rotation period as it is to calculate exactly the yield of forested land. There is therefore both a desire and a need to establish a scientifically founded principle for thinning. The question thereby arises of whether such a law can be found, and if so, if the available data are sufficient.

Let us first examine whether the propositions that I have in part taken, in part deduced, from Chief Forester Nagel's calculations on woodland value are adequate to these calculations. To facilitate their appraisal I will repeat the individual propositions outlined above in sequence.

First Proposition. The annual increase in value of a wood reaches its maximum when the interval between the trees is 12 times their diameter.

Second Proposition. Excluding the initial years required to form the plant and the very advanced ages, and *assuming a normal interval between trees*, the annual increase in the diameter of the individual trees is an equal and constant figure.

Third Proposition. Given the same restrictions and conditions as in the Second Proposition, the annual growth of timber in the *entire wood* is an equal and constant figure.

The following observations employ these propositions to establish whether the law we seek can be derived from them.

We know from the calculations made in the preceding section on the basis of the second proposition that the growth in physical volume of individual trees is $31(3\delta^2\alpha + 3\delta\alpha^2 + \alpha^3)$ cubic feet. According to the first proposition each tree needs a space of

$$12 \times 12\delta^2 = 144\delta^2.$$

If the woodland area is w sq rods, at the beginning of the year $\dfrac{w}{144\delta^2}$ trees stand in this area. This number multiplied by the growth of individual trees gives the total increase of the wood. In the autumn the diameter of the trees will have increased from δ to $\alpha + \delta$. Each tree then requires $144(\alpha + \delta)^2$ of space. In the entire wood therefore there is room for $\dfrac{w}{144(\alpha + \delta)^2}$ trees. But there are $\dfrac{w}{144\delta^2}$ trees, so to create a normal amount of space

$$\frac{w}{144\delta^2} \div \frac{w}{144(\alpha + \delta)^2}$$ must be cleared away, each tree having a volume of $31(\alpha + \delta)^3$ cubic feet. By comparing this mass of timber with the annual growth we can determine the proportion between growth and the yield from thinning.

There seems therefore to be a possibility that the task we have set can be resolved through calculation if the value of δ is given.

However, application of this formula to trees of varying diameter shows that the proportion between growth and the yield from thinning does not remain the same, for instance is different for $\delta = \frac{1}{2}$ than it is for $\delta = 1$.

This difference arises from the fact that we have treated the space the tree requires in the early part of the year ($144\delta^2$) as sufficient for the entire summer, although the trees, if they are to achieve their maximum growth, need space that grows in step with the growth of their diameter.

For practical purposes consideration of such minutiae might seem tedious; but if we seek to study the laws of nature we need to follow the course of the very smallest detail; and we should not take fright at the prospect of more penetrating and difficult investigations if we are to establish a formula that is not only true for a *given* value of δ but holds for all values of δ and gives the correct proportion between growth and the yield from thinnings.

During the growing period in summer the diameter of a tree gains in girth every day, every hour, indeed at every moment. A tree therefore needs more space on each succeeding day than it had the preceding day. If we now calculate the space that a tree requires on each individual day, and then add the space for the individual days and divide by the number of days we will be able to determine fairly precisely the space that a tree has needed during the summer. The result is even more exact if we take the hours as a standard measure and if we were able to determine the increase in diameter moment by moment our calculations would then enjoy a complete agreement with reality.

We can illustrate this gradually increasing need for space as follows, dividing the summer into ten phases. If the diameter = 1 then:

Phase of Summer	Diameter	Space Needed by the Tree
1	$1 + \frac{1}{10}\alpha$	$(1 + \frac{2}{10}\alpha + \frac{1}{100}\alpha^2)144$
2	$1 + \frac{2}{10}\alpha$	$(1 + \frac{4}{10}\alpha + \frac{4}{100}\alpha^2)144$
3	$1 + \frac{3}{10}\alpha$	$(1 + \frac{6}{10}\alpha + \frac{9}{100}\alpha^2)144$
4	$1 + \frac{4}{10}\alpha$	$(1 + \frac{8}{10}\alpha + \frac{16}{10}\alpha^2)144$
5	$1 + \frac{5}{10}\alpha$	$(1 + \frac{10}{10}\alpha + \frac{25}{100}\alpha^2)144$
6	$1 + \frac{6}{10}\alpha$	$(1 + \frac{12}{10}\alpha + \frac{36}{10}\alpha^2)144$

7	$1 + \frac{7}{10}\alpha$	$(1 + \frac{14}{10}\alpha + \frac{49}{10}\alpha^2)144$
8	$1 + \frac{8}{10}\alpha$	$(1 + \frac{16}{10}\alpha + \frac{64}{10}\alpha^2)144$
9	$1 + \frac{9}{10}\alpha$	$(1 + \frac{18}{10}\alpha + \frac{81}{10}\alpha^2)144$
10	$1 + \frac{10}{10}\alpha$	$(1 + \frac{20}{10}\alpha + \frac{100}{100}\alpha^2)144$
	Sum	$(10 + 11\alpha + 3.85\alpha^2)144$

Dividing this sum by 10, the number of phases, gives the average space that the tree requires in summer:

$$(1 + 1.1\alpha + 0.385\alpha^2)144$$

If we now consider the ordering of the coefficients of α and α^2 we see that the sum of coefficients of α is equal to the sum of natural numbers (1, 2, 3 etc.) multiplied by 2 and divided by 10.

The sum of coefficients of α^2 is by contrast equal to the sum of the squares of the natural numbers 1, 2, 3 etc.

If we divide the summer into n phases, the sum of the coefficients of

$$\alpha = (1 + 2 + 3 + \dots \quad n)\frac{2}{n} = \frac{n(n+1)}{2} \times \frac{2}{n} = n + 1$$

The sum of coefficients of $\alpha^2 = \left(1 + 4 + 9 \dots + n^2\right)\frac{1}{n^2}$

The sum of the squares of the natural numbers from 1 to n is however $\frac{2n^3 + 3n^2 + n}{6}$, hence the sum of the coefficients of

$$\alpha^2 = \left(\frac{2n^3 + 3n^2 + n}{6}\right)\frac{1}{n^2} = \frac{1}{3}n + \frac{1}{2} + \frac{1}{6n} \text{ or } \frac{2n^2 + 3n + 1}{6n}.$$

We have therefore estimated the sum of the space that the tree requires in n phases; and so to now estimate the space that the tree requires in a summer of n phases we have to divide the sum we have found by n.

For the space on the ground that a tree of 1 foot diameter requires we have the expression

$$\left(1 + \frac{n+1}{n}\alpha + \left(\frac{1}{3} + \frac{1}{2n} + \frac{1}{6n^2}\right)\alpha^2\right)144 \text{ sq.rods}$$

If we set the diameter of the tree early in the year $= \delta$ instead of 1, then the above calculation of the space required by a tree of δ diameter gives:

$$\left(\delta^2 + \frac{n+1}{n}\alpha\delta + \left(\frac{1}{3} + \frac{1}{2n} + \frac{1}{6n^2}\right)\alpha^2\right)144$$

Let us now consider the emergence of the coefficient of α more closely.

We found this for the individual phases, adding the last term n to the first term 1 in the series $(1,2,3....n)\frac{2}{n}$, and then multiplied this sum first by half the number of terms and then by the second factor $\frac{2}{n}$.

Now n is an indefinite number that we can make as large as we wish. The size of the phases that we assume to be the first term of the arithmetic series and add to the final term following the above procedure is therefore entirely dependent on the size of n, and for instance if $n = 1,000$ the phase lasts for one-thousandth part of a summer.

This period of time has been created by ourselves and does not arise naturally, since the thousandth part of a period such as a summer itself arises from earlier periods of time.

To recognize the law that nature here follows we cannot employ any arbitrary period of time selected by ourselves and add it to the final term. We have rather to add a natural occurrence to the final term. This is not a quantity, but a quality that cannot be added. A dimension first arises when we divide up the flow of becoming.

Hence we have to treat the first term, insofar as it is a quality, as 0, and if we do so then the coefficient of α is transformed from $\dfrac{n+1}{n}$ into $\dfrac{n+0}{n} = 1$.

The same result is obtained if, as happens in such cases in the analysis of infinite series, n is taken to be infinitely large, hence that $\dfrac{1}{n}$ is an infinitely small number; then in the coefficient of $\alpha = \dfrac{n+1}{n} = \dfrac{n}{n} + \dfrac{1}{n}$ the infinitely small unit $\dfrac{1}{n}$ can in addition be treated as equal to 0 (but not in multiplication) – and the coefficient then becomes equal to $\dfrac{n}{n} = 1$.

If $n = \infty$ then in the coefficient of $\alpha^2 = \dfrac{1}{3} + \dfrac{1}{2n} + \dfrac{1}{6n^2}$ the second and third terms $= 0$ and the coefficient is therefore ⅓.

Accordingly, the amount of ground that a tree whose diameter early in the year $= \delta$, during the entire summer $= (\delta^2 + \alpha\delta + \frac{1}{3}\alpha^2)144$.

Having established these propositions we can now turn to calculation of the yield from thinning.

Since we assume that during the summer the trees always have a normal amount of space, it follows that thinning has to progress in step with the increase in diameter of the trees.

1. Growth of the individual trees.

The diameter of each tree is initially

In spring	$= \delta$;
In autumn	$= \alpha + \delta$;

The volume of the tree is

In spring $\qquad = 31\delta^3$

In autumn $\qquad = 31(\delta + \alpha)^3$

$\qquad\qquad\qquad = 31(\delta^3 + 3\alpha\delta^2 + 3\alpha^2\delta + \alpha^3)$

If one subtracts from this the volume of the tree in the early part of the year $-31\delta^3$, then we have a growth of $\quad = 31(3\alpha\delta^2 + 3\alpha^2\delta + \alpha^3)$.

2. Growth of the whole wooded area.

As we have established, the tree requires on average during the whole summer a space amounting to

$$144(\delta^2 + \alpha\delta + \tfrac{1}{3}\alpha^2).$$

Hence on the area of woodland w there can stand

$$\dfrac{w}{144\left(\delta^2 + \alpha\delta + \dfrac{1}{3}\alpha^2\right)};$$ the growth of each individual tree

is $31(3\alpha\delta^2 + 3\alpha^2\delta + \alpha^3)$. The total growth of all trees is therefore

$$\dfrac{31w\left(3\alpha\delta^2 + 3\alpha^2\delta + \alpha^3\right)}{144\left(\delta^2 + \alpha\delta + \dfrac{1}{3}\alpha^2\right)}$$

$$= \dfrac{31w}{144} \times 3\alpha.$$

Example. If w = 144,000 square feet = 562.5 sq. rods, $\alpha = \frac{1}{6}$ inch = $\frac{1}{72}$ of a foot, then the growth of the wood

$$= 31 \times \dfrac{144,000}{144} \times \dfrac{3}{72} = 31,000 \times \dfrac{3}{72} = 1,292 \text{ cu.ft.}$$

3. Growth of remaining stock.

At the beginning of spring each tree has a diameter of δ, a volume of 31 δ^3 and requires a space of 144 δ^2.

Therefore on the woodland area $w\dfrac{w}{144\delta^2}$ trees can stand.

The timber stock of the wood is therefore

$$\frac{w}{144\delta^2} \times 31\delta^3 = \frac{31w\delta}{144}.$$

In the autumn at the end of the growth period the diameter of the tree $= \delta + \alpha$, the volume $= 31(\delta + \alpha)^3$, and the space occupied by the tree $= 144(\delta + \alpha)^2$. On the area w therefore $\dfrac{w}{144(\delta + \alpha)^2}$ trees can stand.

Multiplying the number of trees by the volume of the individual trees gives

$$\frac{w}{144(\delta + \alpha)^2} \times 31(\delta + \alpha)^3 = \frac{31w(\delta\alpha)}{144}.$$

Subtracting the stock at the beginning of spring $\dfrac{31w\delta}{144}$ gives the growth of the remaining stock of timber as

$$\frac{31w(\delta + \alpha)}{144} - \frac{31w\delta}{144} = \frac{31w}{144} \times \alpha.$$

Total growth we have found to be $\dfrac{31w}{144} \times 3\alpha$. In the autumn only $\dfrac{31w}{144} \times \alpha$ remains. Thinning has therefore removed

$$\frac{31w}{144} \times 3\alpha - \frac{31w}{144} \times \alpha = \frac{31w}{144} \times 2\alpha.$$

We arrive at the extremely remarkable result that *given the normal situation of the trees, the yield from thinning in volume of wood is twice the amount of growth of the remaining stock.*

Example: if δ = 1 foot, α = $\frac{1}{72}$ feet, w = 1,440,000 square feet = 5,625 sq. rods.

The space needed by a tree for the entire summer is $(\delta^2 + \alpha\delta + \frac{1}{3}\alpha^2)144$, and for these values is therefore 144 × 1.014 sq. rods.

On this area of woodland there can therefore stand

$$\frac{1,440,000}{144 \times 1.014} = 9,862 \text{ trees.}$$

The tree which has early in the year a diameter of 1 foot grows in the course of the summer to $1\frac{1}{72}$ feet in diameter and its volume grows from 31 × 1^3 = 31 cubic feet to $31(\frac{73}{72})^3 = 32.31$ cubic feet.

The growth of the tree is therefore 32.31 − 31 = 1.31 cubic feet. From this the total growth of 9,862 trees which during the entire summer enjoy a normal amount of space = 9,862 × 1.31 = 12,919 cubic feet.

In the course of the summer constant thinning removes as many trees as required for the creation of a normal amount of space. In the autumn the normal space for a tree $1\frac{1}{72}$ feet in diameter = $144\left(\frac{73}{72}\right)^2 = 144 \times \frac{5,329}{5,184}$.
Hence on the entire woodland area there stand

$$\frac{1,440,000}{144 \times \dfrac{5,329}{5,184}} = 9,728 \text{ trees.}$$

The volume of a tree is in
the autumn   =   32.31 cubic feet.
The remaining stock is
therefore 9,728 × 32.31   = 314,311 cubic feet.

Earlier in the year the
stock had been   = 310,000 cubic feet;
hence the growth of
the remaining stock is   =   4,311 cubic feet,
while the total growth is   = 12,919 cubic feet.
Thinning therefore removes 8,608 cubic feet.

(The fact that thinning has not exactly removed twice the increase of the remaining stock is due entirely to calculating in decimals).

*   *   *

It is easier to appreciate the required calculation if one assumes that thinning is not carried out constantly throughout the summer, but is done once only, in the autumn. Hence if there are 10,000 trees on the woodland area in autumn, each of 32.31 cubic feet then

The entire timber stock   = 323,100 cubic feet
The stock early in the year = 310,000 cubic feet
Total growth is therefore   13,100 cubic feet.

If the trees now are given a normal amount of space through thinning, then there is space only for 9,728 trees (as already calculated), hence 272 trees have to be removed. The yield from thinning is therefore
272×32.31 =   8,789 cubic feet.
Total growth   = 13,100 cubic feet,
from which   4,311 cubic feet
remains in the wood.

The growth in the remaining stock relates to the yield from thinning in the proportion

4,311 : 8,789 = 100 : 204.

Here the yield from thinning is over $\frac{2}{3}$ of the total growth.

This calculation does however lack mathematical precision. For if the trees only have a normal amount of space early in the year, during the course of the summer they become increasingly crowded, and cannot achieve the growth calculated here. This form of calculation also gives for trees of varying diameter a different proportion between the remaining growth and the yield from thinning. Likewise, it is not possible to determine from this the law of nature that prevails here, although the proposition derived from the above calculated formulae does have a general validity.

If one instead takes as a basis the space that the tree needs in midsummer rather than in spring, then the deviations between the results of the two forms of calculation is insignificant.

§ 17.
Criticism.

Most of the material in the preceding § was worked through and written up in 1828.

The result of this study was so surprising to me and stood is such stark contrast to the practice and views of experienced foresters that I suspected I had made an error in the reasoning, and consequently disregarded the results themselves.

Since I did not at that time have the leisure to work through it once more, I put the study aside and almost entirely forgot about it. More recently, however, I was promoted to clarify for myself the value of timber stocks of different age, and also the rent produced by using land for the cultivation of pines. The ground rent depends largely on the rotation period and the yield from thinning. My continued study quickly showed that the most advantageous rotation period is itself dependent on the yield from thinning – therefore this became the pivot of the investigation.

However, since I lack data on the yield from thinning, and know of no authority on the matter whom I can follow, I had perforce to make several hypothetical assumptions in arriving at propositions relating the yield from thinning to the remaining stock, which would then enable me to approach more exactly the truth of the impact of different methods of thinning on the rotation period and ground rent.

It was in this way that the first section above came about. But the deeper I went into the matter, the more I found new questions that could not be resolved by hypothetical propositions, and every step revealed more clearly the need for a knowledge of the law that governed this in nature.

I was necessarily drawn back in this way to my earlier study from 1828. I have carefully re-examined what I then wrote; but I cannot discover any error in the reasoning itself.

I have therefore to leave it to others to try and find fault with its argument and conclusion.

If this is unsuccessful, criticism will have to turn upon Nagel's propositions, which, where actual figures, rather than algebraic variables, are involved, forms the basis for our own work; although I cannot of course testify to the accuracy of these figures. Only by such criticism can it be determined whether the result established here, or prevailing opinion, is a falsehood.

To this end we turn now to the question of whether, and to what extent, modification of Nagel's propositions has an influence on the main finding of our study: that given a normal relative interval between trees, the timber yielded from thinning amounts to $\frac{2}{3}$ of the total growth.

Nagel's first proposition, according to which the trees should be at a mutual interval 12 times as great as the diameter of their trunks if timber growth was to be maximized – certainly deserves repeated examination.

Accordingly, we need to examine whether there is an interval other than 12 times the diameter that compels this law to be altered.

If we make r rather than 12 the interval and $r^2\delta^2$ instead of 144 δ^2 the normal amount of space, the total amount of growth becomes $\dfrac{31w}{r^2} \times 3\alpha$; and for the growth of the remaining stock $\dfrac{31w}{r^2} \times \alpha$. That part of the total growth that does not enter into the remaining stock is the yield of thinning, which makes up $\dfrac{31w}{r^2} \times 2\alpha$.

We therefore once more gain the result that thinning removes $\frac{2}{3}$ of the growth, or double the growth that remains in the stock. Since r can be 10 or 15 or any other number corresponding to the requirement that the wood achieves the greatest possible growth, the law established above is unaffected by whatever figure one might later find to be the correct one regarding the interval between trees.

From the second proposition of Nagel, that the increase in the diameter of the trees follows an arithmetic series can be deduced the third, that the increase of timber stocks in the entire wooded area occurs in an arithmetical progression, as the following shows.

If the regular annual growth of trees in respect of diameter = α, the normal interval = r, and that the diameter of trees in the spring of the nth year of their age = δ, then, as shown in the section above, on the woodland area w there stand $\dfrac{w}{r^2\delta^2}$ The volume of each tree is (§ 15) $= 31\delta^3$, and the timber stock is

$$\frac{w}{r^2\,\delta^2}\times 31\delta^3 = \frac{31w\delta}{r^2}.$$

Early in year $n + 1$ the trees have a diameter of $\delta + \alpha$ and on area w there can stand $\dfrac{w}{r^2(\delta+\alpha)^2}$; the volume of each tree is then $31(\delta+\alpha)^3$, and the wood stock is

$$\frac{w}{r^2(\delta+\alpha)^2}\times 31(\delta+\alpha)^2 = \frac{31(\delta+\alpha)}{r^2}.$$ In the spring of the

nth year the stock was $= \dfrac{31w\delta}{r^2}$; subtracting this stock leaves the growth for the nth year (as in the previous section) $\dfrac{31w}{r^2}\times\alpha$.

Early in the year $n + 2$ the diameter of the tree is $\delta + 2\alpha$, the volume of the tree is $31(\delta + 2\alpha)^{3.}$ On the woodland area w $\dfrac{w}{r^2(\delta+2\alpha)^2}$ trees can stand.

The timber stock is thus $\dfrac{w}{r^2(\delta+2\alpha)^2}\times 31(\delta+2\alpha)^3 =$

$\dfrac{31w(\delta+2\alpha)}{r^2}$. Subtracting from this the stock of year

$n + 1$ of $\dfrac{31w}{r^2}(\delta+\alpha)$ gives for the year $n + 1$ a growth of $\dfrac{31w}{r^2}\times\alpha$.

The growth in year $n + 1$ is there just as great as in year n, and it can easily be seen that continuing the calculation for the years $n + 2$, $n + 3$ etc. will always lead to the same result.

The remaining timber stock of the woodland area increases therefore in direct proportion to the increase in diameter of the individual trees.

Nagel's proposition that the remaining timber stock increases according to an arithmetic progression would be contradicted by many foresters, pointing to the fact that this proposition is not confirmed in practice.

Chief Forester Nagel claims, however, that where experience does not confirm this proposition, mistakes have been made in the thinnings.

Who then is right?

Let us make a comparison with details supplied by the honourable veteran of silviculture, Chief Forester Cotta.

Cotta provides in his tables on the determination of volume and growth of complete stocks (p. 41), the stock for pine afforestation of the tenth class, that is the best Saxon arable land, the following figures:

Age of tree in year	Wood stock in cubic feet	Growth in cubic feet
20	2,940	
21	3,123	183
30	4,850	
31	5,053	203
40	6,950	
41	7,166	216
50	9,150	
51	9,374	224
60	11,350	
61	11,564	214
70	13,450	
71	13,654	204
80	15,450	
81	15,644	194

90	17,350	
91	17,534	184
100	19,150	
101	19,324	174
110	20,820	
111	20,978	158
120	22,320	
121	22,460	140

Here the growth of trees rises in the early years, and then later falls.

Cotta himself admits (*Waldbau*, 4th edition, p. 106)[2] that he took a very long time to arrive at proper rules for thinning, and apply them. It is therefore very probable that the observations upon which his tables for yield are based are founded upon stocks that have been deficiently thinned. This probability increases for the older stocks, since the propositions of thinning came to the fore relatively late in the day, and there are perhaps no older stocks that have been properly treated since their early years.

While the deviations of Cotta's tables from Nagel's propositions cannot therefore be considered evidence against the latter, this does not prove that Nagel's propositions are correct.

We therefore have to investigate whether, and to what extent, the result of our study depends on these underlying propositions.

For this purpose we assume – as with the Cotta tables – that growth initially rises, reaches its highest point and thenceforth decreases once more.

---

2. Editorial addition: Heinrich Cotta, *Anweisung zum Waldbau*, Dresden und Leipzig 1817.

For the age of trees of *n* years the increase in diameter = α. If this age falls in the period during which growth is increasing, the increase of diameter is greater than α. If however the year n falls in the period during which growth decreases, then the increase in diameter is smaller than α, or = $(1 - t)\alpha$. Both cases are captured in the expression $(1 \pm t)\alpha$.

If we substitute $(1 \pm t)\alpha$ for α, and $r^2\delta^2$ for $144\,\delta^2$ in the calculation of incremental growth given in the preceding paragraphs, we obtain the following results:

1) Total growth is $= \dfrac{31w}{r^2}(1 \pm t)3\alpha$.

2) The growth of the remaining stock is $= \dfrac{31w}{r^2}(1 \pm t)\alpha$.

The relation between the two is therefore

$$\frac{31w}{r^2}(1 \pm t)3\alpha : \frac{31w}{r^2}(1 \pm t)\alpha$$

$$= 3 : 1.$$

The relationship therefore remains exactly the same as it was in the preceding §, which proves that the rectitude of the law – according to which with normal thinning the yield of such thinning is double the growth of the remaining stock – is independent of the rectitude of Nagel's propositions.

* * *

From this angle the result of our investigation cannot be assailed. It can however be argued that the space that the trees require, if the entire woodland area is to deliver its greatest increase, has *absolutely no relation* to the diameter of the trees; and in this case the entire argument collapses.

But the standard for the normal amount of space can only be derived from the tree itself, and if the diameter is not used, there is nothing but the volume of the tree itself than can be used.

We will now assume as a hypothesis that the space needed by every tree, if the wooded area is to achieve the greatest increase, stands in direct relationship to the timber volume of the tree.

The tree that contains $31\delta^3$ cubic feet requires a space of r^2 square feet.

Each cubic foot of timber mass therefore enjoys a space of $\dfrac{r^2 \text{ sq. ft.}}{31\delta^3}$.

During a summer the diameter of the tree increases from δ to $\delta + \alpha$ and the volume of the same from $31\delta^3$ to $31(\delta + \alpha)^3$. Since, however, one cubic foot of timber mass requires $\dfrac{r^2 \text{ sq. ft.}}{31\delta^3}$ of space, the tree whose volume is $31(\delta + \alpha)^3$ cubic feet requires a space of $31(\delta + \alpha)^3 \times \dfrac{r^2}{31\delta^3}$.

On a wooded area w therefore there can stand

$$w : \left(31(\delta + \alpha)^3 \times \frac{r^2}{31\delta^3} \right) = \frac{w\delta^3}{r^2 (\delta + \alpha)^3} \text{ trees.}$$

Each tree has a volume of $31(\delta + \alpha)^3$, the timber volume of the wood is therefore $\dfrac{31w\delta^3}{r^2}$.

If one compares this with the timber volume earlier in the year, then we derive the magnitude of the increase of the remaining stock.

Early in the year the volume of the tree is $31\delta^3$; the space that the tree requires $= r^2$ square feet. On an area w therefore

$\dfrac{w}{r^2}$ trees can stand. The timber volume of the wood is therefore $\dfrac{w}{r^2} \times 31\delta^3 = \dfrac{31w\delta^3}{r^2}$.

For the following year we have found that the timber volume is $\dfrac{31w\delta^3}{r^2}$, hence exactly as large as in the preceding spring.

From this it follows that if the timber volume of the tree is the standard for the space that the tree requires, the *total* growth in every year has to be removed by thinning and there is no increase in the timber volume of the wood, which is an absurdity.

Example: if $\delta = 1$, $\alpha = \frac{1}{72}$, $r = 12$ feet and $w = 1{,}440{,}000$ sq. ft.

The volume of the tree is $31\delta^3 = 31$ cubic feet. The tree requires a space of 144 sq. ft. One cubic foot of timber mass requires therefore $\frac{144}{31} = 4.645$ sq. ft.

A tree of 1 foot in diameter early in the year is in the autumn $1\frac{1}{72}$ feet in diameter, and has a volume of $31(\frac{73}{72})^3 = 32.31$ cubic feet. The space which such a tree requires is thus $32.31 \times 4.645 = 150.1$ sq. ft. On the area w of $1{,}440{,}000$ sq. ft. therefore $\dfrac{1{,}440{,}000}{150.1} = 9{,}595$ trees can stand.

The timber volume in the wood is
then $9{,}595 \times 32.31$ = 310,014 cubic feet.
In spring the timber volume was
$10{,}000 \times 31$ cubic feet = 310,000 cubic feet.

The timber volume is therefore equally large in autumn and in spring, which presupposes thinning that removes all growth.

It can generally be derived that, if the normal space needed by a tree is related in all years to its timber volume, the entire growth of the wood has to be removed in thinning and the timber stock of the wood will always be the same.

If, on the other hand, one takes as the normal stock the timber stock of a wood containing trees of a specific size, for example one foot in diameter, then in later years there will only be growth in the remaining trees which would be equal to the timber mass of those trees that have been removed. But the younger wooded areas would then everywhere have the normal timber stock, and every successive removal of trees would move away even further from the normal stock; and so younger stock should not be thinned at all, leaving them prey to all the tribulations of wastage that arises with the struggle of trees for space.

Continuing these hypotheses to such absurdities and contradictions demonstrates their irrelevance.

Since the standard for the space required by a tree has to lie in the tree itself, there is no alternative but to return to the assumption that the diameter of the tree is the standard for the normal interval and the need for space.

* * *

If one takes the proposition that thinnings must remove $\frac{2}{3}$ of the growth as the axiom of silviculture, the mere practitioner will say:

> What use to me is the knowledge of a law that depends entirely upon a series of assumptions that can never be found in reality? Here trees of equal diameter and equal growth are assumed, all of which are perfectly and equally distributed through a wood, without the slightest clearing; constant and steady thinning is also assumed, whereas in practice thinning can be done

only after periods of several years. What use can I make of studies which, like these, are based upon entirely ideal circumstances?

To which I reply:

These investigations directed to the natural laws governing forestry relate to practical silviculture as pure geometry relates to applied.

Pure geometry is based upon entire fictions: points without extension, lines of no width – none of these can be found in reality. Nonetheless, pure geometry is the irrefutable foundation of practical geometry, and without the former the latter would be nothing but trial and error.

§ 18.
Ground rent and the most advantageous rotation period, if only $\frac{1}{3}$ of the growth is left in the remaining stock.

In this case $\frac{2}{3}$ of the annual growth, that is $\frac{2}{3} \times 150 =$ 100 cubic feet, is removed in thinning. Timber stocks of five- to ten-years-old are excluded from this, since the annual growth is here 100 instead of 150 cubic feet per year, 50 cubic feet remaining for the stock and 50 cubic feet being removed by thinning.

According to § 3, if the annual increase of the timber stock is 100 cubic feet per *Waldmorgen*, the timber value of $(x + 5)$ year-old stock is $\qquad 3x^2 + 15x;$

and the timber value of all stock from
years 1 to 5 is $\qquad x^3 + 9x^2 + 8x.$

If however, as here, the annual growth of stock is only 50 cubic feet per year, and not 100, then both values are reduced by one half.

The yield from thinning is given by the following calculation.

a. When the wood removed in thinning has the same value per cubic foot as the remaining stock.

With the exception of the felling of five- to ten-year-old trees, the woodland yields 100 cubic feet of wood from thinning per *Waldmorgen*.

In § 3 we found that for a growth of 100 cubic feet per year in the timber stock the value of the timber removed is $3x^2 + 15x$.

If we examine the stocks from the 1st to the $(x + 5)$th year we can see that the value of clear felling can be derived from the sum of the increase in value for all stocks.

If every stock gives an annual thinning yield of 100 cubic feet, that is, as much as the increase of the stock, then the value of all thinning must equal the value of the clear felling, hence $3x^2 + 15x$. But since the stock from the 6th to the 10th year give a lesser increase than that for the older trees, and the yield from these thinnings is not 100 cubic feet but 50; it is necessary to subtract half of the value of the ten-year-old stock, which is (as in § 3) 150.

Accordingly, the yield from thinning is $3x^2 + 15x - 75$.

b. When the wood removed in thinning is only $\frac{2}{3}$ the value per cubic foot of the remaining stock.

Since our investigation is based on this assumption, we need to subtract from the yield calculated in a., and so we get for the value of all thinnings the expression

$$\frac{2}{3}(3x^2 + 15x - 75)$$
$$= 2x^2 + 10x - 50.$$

We are now able to arrive at a general formula for the magnitude of ground rent in cases where

1) the total growth is 150 cubic feet,

2) the remaining stock increases annually by 50 cubic feet,

3) and thinning gives 100 cubic feet per *Waldmorgen.*

Revenue:

1) From clear felling	$1.5x^2 + 7.5x$
2) from thinning	$2x^2 + 10x - 50$
Together	$3.5x^2 + 17.5x - 50.$

Expenditure:

1) Interest on the value of all timber stocks
   $= 0.5x^3 + 4.5x^2 + 4x$
   at 4% which is $0.02x^3 + 0.18x^2 + 0.16x$
2) Seeding cost for
   130 sq. rods at 2 sh.   260
3) Administration and
   supervision costs for $x + 5$
   Waldmorgen at 8 sh. $8.0x + 40$

Total expenses $0.02x^3 + 0.18x^2 + 8.16x + 300$

Subtracting these expenses
from the revenue gives
the ground rent of $-0.02x^3 + 3.32x^2 + 9.34x - 350.$

If we give the values 50, 60, 70 etc. for x, we get the following results:

For a rotation of	The ground rent per *Waldmorgen* is
55 Years	107.6 sh.
65 Years	120.6 sh.
75 Years	129.5 sh.
85 Years	134.2 sh.
93 Years	135.0 sh.
95 Years	134.7 sh.
105 Years	131.3 sh.

From this it is shown that the highest ground rent results from a rotation period of 93 years.

Comparison:

Annual amount of growth removed by thinning	Most advantageous rotation period (in years)	Ground rent per *Waldmorgen* (sh.)
One-third	55	78.8
One-half	67	97.1
Two-thirds	93	195.0

This shows quite strikingly how the net yield of woodland increases the more vigorous the thinning procedure, the longer the most advantageous rotation period, and makes the production of stronger building timber profitable.

*   *   *

Hitherto in all our studies we have considered only one aspect of thinning, that is, its influence on ground rent given a *constant rate of increase of the woodland timber.*

However, the greater or lesser amount of cleared area resulting from stronger or weaker thinning has an effect not only on the growth in diameter of the individual trees but has the same effect upon the total growth of the wood.

If we fail to include this effect of thinning upon growth in our calculations the solution to the task we have set ourselves will remain incomplete. The search for such a solution leads us into a new and more complex investigation.

Section Three.

§ 19.
What is the relationship between the growth of the tree and the space which each tree is given?

To answer this question we have first of all to deal with the space that a tree of a given diameter requires for its roots.

Several years ago, while making an entrance to a meadow, soil was removed close to some willows (*Salix alba*) lining a road, and I noticed that the roots of the willows reached up to 4 rods from the line of trees. I have also been told of finer roots at a distance of 6 rods from the trees. On the other side of the line of trees there is a country road separated from the trees by a ditch. The trees themselves, which are about 9 inches in diameter, are 16 feet apart. If we assume that the trees have not grown any roots in the direction of the road, the area used by each tree is $= 1 \times 4 = 4$ sq. rods $= 1,024$ sq. ft. The square of the diameter, or δ^2, is here $\frac{3}{4} \times \frac{3}{4} = \frac{9}{16}$ sq. ft. The area penetrated by the roots is $1,024 : \frac{9}{16} = 1,820 \, \delta^2$.

If the willows were not pollarded every three years the trunks would probably be significantly thicker, but the spread of the roots would probably be no greater.

Root spread varies, however, between different types of tree, pines and birches having a smaller spread than the willow, while the Canadian poplar has an even greater spread than the willow.

I have made the following observations in respect of the birch.

Here in the garden there is a pathway 12 feet wide lined on either side with birches spaced at intervals of 16 feet, the trees having an average diameter of 7½ inches = $\frac{5}{8}$ feet. Beyond the pathway there is a piece of land planted with beet. On a strip about 16 feet wide the beet does very poorly, the remainder, however, doing much better. This rough distinction can be treated as proof that the spread of roots from the trees reaches 16 feet, and no further. The area of land that the tree uses is therefore $(6 + 16)16 = 22 \times 16 = 352$ sq. ft. The square of the diameter of the trees is $\delta^2 = \frac{5}{8} \times \frac{5}{8} = \frac{25}{64}$ sq. ft. The area over which the tree has spread its roots is accordingly $352 : \frac{25}{64} = 901\delta^2$. If the trees were evenly and squarely spaced the distance from one tree to the next would be $\sqrt{901\delta^2} = 30\delta$. Hence birch trees whose distance from each other is 30 times their diameter would spread their roots across the entire area of the wood and use it for the production of timber.

I have not had the opportunity of observing the root spread of isolated pine trees, and for lack of data assume that this is the same for both birch and pine.

The growth of isolated trees is significantly greater than that of trees grouped in stands, and we can clearly observe that with the reduction in relative space the growth of trees decreases. If the interval between the trees is constantly reduced in relation to their diameter, there comes a point where their growth entirely ceases. The point where growth becomes equal to zero arises much sooner than the point where the interval between the trees becomes zero.

The space occupied by the tree where its growth ceases owing to crowding, and becomes zero, I call the *space required to sustain the tree*.

Such a reduction of relative space occurs in a young wood where the size of the trunks increases but no trees are removed.

The absolute space taken up by each tree then remains the same, but the space in relation to the diameter decreases.

If we now presume a wood with an entirely level and uniform soil, in which all trees are of the same size, health and vigour, and are placed at the same interval, then with the further growth of the trees, and in the absence of any thinning, ultimately a point comes where the trees have only the space necessary to sustain them.

Assuming then that if the tree ceases growth it means that the tree dies, then not only individual trees, but the entire wood dies simultaneously.

If anyone doubted this and objected that

some trees would die off, and this would create space for other trees, and so, as we see everywhere in reality, not all trees would simultaneously die.

Then one would have to ask which trees would die and which survive, and since the position, interval, health and vigour of all these trees are the same, it would not be possible to explain why this tree should die, while an entirely similar tree survived.

We know for mature animals that, if their feed is reduced to a certain amount, they cease to put on weight but instead continue in the same state in which they find themselves. If their feed is increased, then they begin to put on weight; and this increase can be attributed entirely to the additional feed, since the portion that they had received before sufficed only to maintain them in existence.

If one gives younger, immature animals – calves, piglets – an amount of feed entirely adequate to sating animals of that particular age, they put on a considerable amount of weight. As the body grows their need for feed also increases. If one continues to give them the earlier amount of feed, the animal's growth gradually slows and finally ceases altogether. The same quantity of feed that sates a young animal and that sufficed for its growth is just enough simply to sustain the older and larger animal. The growth of a young animal is quite natural, and it is probable that if this growth were stopped through insufficient feed then it would not simply cease growing, like the mature animal, but die.

Since it is undeniable that at a certain relative distance of one tree from another the growth of all these trees ceases, and the nourishment that the trees draw from the soil remaining to them provides nothing for their growth, but is used simply to sustain them: thus we encounter here an unmistakable analogy between the animal world and the world of plants.

If we give for the interval at which trees cease growth = $m\delta$, the area that the tree then uses = $m^2\delta^2$; and if the trees are given a space of $(m + 1)\delta$ through thinning and start to grow again, then this growth appears as the product not of the entire space occupied by the tree, but rather of the additional space that the tree has been given.

The amount of feed required to sustain an animal is governed by its body mass, for a tree by the square of its diameter. Here there appears to be a difference between the two. But when one takes into account that the spread of the roots is not only horizontal, but vertical, it can be seen that the amount of space required to sustain a tree is likewise in proportion to its mass.

Let us continue with this comparison. If a cow of 600 lb bodyweight has a daily need of 12 pounds of hay or its equivalent to sustain it, and is completely sated by 20 pounds of hay, one can assume that the incremental increase to 13, 14, 15, 16, 17, 18, 19, and 20 pounds will bring about an equal increase in animal products, such as milk and meat.

Here there is a significant difference between animals and trees.

The additional feed for the animal is placed directly by it and so the animal is able to completely consume it.

The tree in the wood, by contrast, does not find any nutritional increase in the immediate vicinity of the trunk, but instead in the increased space made available to it for the spread of its roots, and the tree is not able to make the same use of distant nutrition as it can of that closer to the trunk.

The main roots of the tree spread themselves if not in straight lines then in a ray, becoming more distant from each other the further they extend from the trunk; and the sucking roots are not able to penetrate the earth so effectively, nor gather the nutrition in the soil, as happens closer to the trunk.

If one were to draw a series of concentric circles around the trunk representing its growth, the volume of these concentric rings increases the further they are from the trunk, and consequently they contain an ever-greater amount of plant nutrition. On the other hand, as the interval increases, the capacity of the tree to extract this nutrition constantly diminishes – and the question arises of the law according to which this diminution occurs.

If we had three completely reliable observations regarding the rate of growth of trees placed at, for instance, relative intervals of 8, 12 and 16 times their diameter from each other it would be possible to derive such a law.

In the absence of such observations we have to be content with hypothetical assumptions whose rectitude can be checked only by the results that they lead to.

Accordingly, I will put forward the following proposition:

If the trees standing in a square require for their basic existence an interval of $m\delta$, and that with an interval of $(m + 1)\delta$ the growth of their diameter $= g$, then with an interval of $(m + 2)\delta$ the growth is $2g$, with an interval of $(m + 3)\delta$ the growth is $3g$ and so forth, until the capacity of the tree to extend its roots and make use of the room available finds its limit.

Example. If the interval necessary to maintain the life of the tree $m\delta = 6\delta$, the diameter $\delta = 1$ foot, then if at an interval of 7 feet the growth is 1 line, the growth for the interval

Of	8 feet is	2 lines
Of	9 feet is	3 lines
Of	10 feet is	4 lines
Of	11 feet is	5 lines
Of	12 feet is	6 lines, etc.

The proposition is made more probable by the observation that under isolated trees garden plants and field crops do worst near to the trunk, improving the more removed they are from the trunk.

If growth entirely ceases with a very small interval between the trees, while with a very great interval some part of the ground either remains entirely unused, or is of very limited use, then there has to be a particular interval between the trees at which the overall growth of the wood is at its maximum; and the question is where this lies.

To solve this we have to make use of the propositions established in § 15 and § 16. Instead of simply referring to these paragraphs, I will give a brief summary again here.

The volume of a tree of diameter δ is $31\delta^3$. For the interval r the growth of the diameter is α. A tree of diameter δ requires a space of $r^2\delta^2$. In a wooded area w therefore $\dfrac{w}{r^2\delta^2}$ such trees can stand.

Each tree has a volume of $31\delta^3$ cubic feet, so the timber stock of the entire wooded area is therefore

$$\frac{w}{r^2\delta^2} \times 31\delta^3 = \frac{2w\delta}{r^2}.$$

During the summer the diameter of the tree increases from δ early in the year by an amount α, so that in the autumn it is $\delta + \alpha$.

The volume of the tree is then $31(\delta + \alpha)^3 = 31(\delta^3 - 3\alpha\delta^2 + 3\alpha^2\delta + \alpha^3)$.

If one subtracts from this the volume of the tree in spring of $31\delta^3$, then the increase of each tree $= 31(3\alpha\delta^2 + 3\alpha^2\delta + \alpha^3)$. The tree whose volume increases over the summer from δ to $\delta + \alpha$ needs on average throughout the summer a space of $r^2(\delta^2 + \alpha\delta + \tfrac{1}{3}\alpha^2)$.

Hence on the wooded area w there can stand $\dfrac{w}{r^2(\delta^2 + \alpha\delta + \tfrac{1}{3}\alpha^2)}$ trees. Multiplying this number of trees by the growth of the individual trees gives for the total growth of the wood

$$\frac{w}{r^2(\delta^2 + \alpha\delta + \tfrac{1}{3}\alpha^2)} \times 31(3\alpha\delta^2 + 3\alpha^2\delta + \alpha^3) = \frac{31w}{r^2}\,3\alpha.$$

Taking ⅔ of this away in thinnings, the amount of growth entering the remaining stock is $\frac{31w}{r^2} \times \alpha$ cubic feet.

How does this growth alter if the interval between the trees alters, becoming y instead of r?

If the interval between trees is r, the portion of that interval bringing about growth, and which one can call the effective interval, we shall call $r - m$. For an interval y the effective interval is $y - m$.

Since however this effective interval is in direct relationship with the diameter of the trees $(r - m) : (y - m) = \alpha : \frac{y - m}{r - m} \times \alpha$.

If we call z the growth derived from interval y, then

$$z = \frac{y - m}{r - m} \times \alpha \cdot$$

For the entire wooded area the growth of the remaining stock will therefore be

$$\frac{31w}{y^2} \times z = \frac{31w}{y^2} \times \frac{(y - m)^\alpha}{r - m} = \frac{31w\alpha}{r - m} \times \frac{y - m}{y^2}.$$

The magnitude of an increase α does not only depend on the interval, but also on the soil and on the site of the trees. For the application of this formula to concrete cases I assume a site in which the annual increase in diameter is ⅙ of an inch or ¹⁄₇₂ of a foot, the trees constantly having a normal amount of space in which their increase in value is at its maximum.

What then is the interval that meets these conditions?

I have no basis for this other than that taken from Nagel's propositions, according to which the interval has to be 12 times the diameter, and so I take r as = 12. The wooded area w can be any area one wishes. Here I take a *Waldmorgen* of 130 square rods = 33,280 square feet. For these values of α, r and w the above formula becomes:

$$\frac{(31 \times 33{,}280 \times \frac{1}{72})}{12 - m} \times \left(\frac{y - m}{y^2}\right) = \left(\frac{14{,}329}{12 - m}\right) \times \left(\frac{y - m}{y^2}\right)$$

In this expression for the growth we find, besides the variable magnitude y the unknown, but constant, magnitude m.

To define the value of m, or the figure for the interval at which growth of the trees ceases entirely, we do however lack any kind of natural observations.

It therefore seems that our effort at determining the relation between interval and growth numerically here runs up against its limit, and we must give up the attempt.

Happily, its value can be identified via analysis of the propositions we have already established.

In the expression for growth $\left(\frac{31w\alpha}{r - m}\right)\left(\frac{y - m}{y^2}\right)$ the first factor is a constant magnitude which remains unchanged after differentiation of the function; in the second factor $\left(\frac{y - m}{y^2}\right)$, y is a variable, and if we take the differential of this and set it equal to zero, then we can find at which value for y growth finds its maximum.

The differential is

$$d\left(\frac{y - m}{y^2}\right) = y^2 dy - (y - m)2y dy = 0$$

$$\text{Hence } y^2 - 2y(y - m) = 0$$

$$y^2 - 2y^2 + 2my = 0$$
$$y^2 = 2my$$
$$y = 2m \text{ and}$$
$$m = \tfrac{1}{2}y.$$

We know from Nagel's propositions that the highest growth occurs when the interval is 12. Since for $y = 2m$ the greatest increase also holds, then $y = 2m = 12$; and so $m = 6$.

If we put this figure in the formula given earlier for growth:

$$\frac{14,329}{12 - m} \times \left(\frac{y - m}{y^2} \right)$$

For $m = 6$ this formula becomes

$$\frac{14,329}{6} \times \left(\frac{y - m}{y^2} \right) = 2,388 \left(\frac{y - 6}{y^2} \right).$$

This formula allows us to calculate growth for whatever intervals we might wish:

If the interval is	The growth per *Waldmorgen* is
$y = 6$	$2,388 \times 0 \quad = \quad 0$ cu. ft.
$y = 7$	$2,388 \times \tfrac{1}{49} \quad = 48.7$ cu. ft.
$y = 8$	$2,388 \times \tfrac{1}{32} \quad = 74.6$ cu. ft.
$y = 9$	$2,388 \times \tfrac{1}{27} \quad = 88.4$ cu. ft.
$y = 10$	$2,388 \times \tfrac{1}{25} \quad = 95.5$ cu. ft.
$y = 11$	$2,388 \times \tfrac{1}{24.2} = 98.7$ cu. ft.
$y = 12$	$2,388 \times \tfrac{1}{24} \quad = 99.5$ cu. ft.
$y = 13$	$2,388 \times \tfrac{7}{169} \quad = 98.9$ cu. ft.
$y = 14$	$2,388 \times \tfrac{2}{49} \quad = 97.5$ cu. ft.
$y = 15$	$2,388 \times \tfrac{1}{25} \quad = 95.5$ cu. ft.

If the interval is	The growth per *Waldmorgen* is
$y = 16$	$2{,}388 \times \frac{5}{128} = 93.3$ cu. ft.
$y = 17$	$2{,}388 \times \frac{11}{289} = 90.9$ cu. ft.
$y = 18$	$2{,}388 \times \frac{1}{27} = 88.4$ cu. ft.
...	...
$y = 24$	$2{,}388 \times \frac{1}{32} = 74.6$ cu. ft.
...	...
$y = 30$	$2{,}388 \times \frac{2}{75} = 63.7$ cu. ft.

At 8 times the interval (always taking the diameter of the tree as a standard) the increase is here only about ¾, at seven times barely half the growth one finds with the normal interval of 12.

This very clearly demonstrates the disadvantages associated with overcrowding stands of timber. But, strangely enough, it appears that humans dislike cutting down living trees – and a close watch has to be kept on forest workers during thinning to see that they do indeed remove enough timber and create sufficient space.

Too great an interval seems here much less a disadvantage for the growth of the entire wood than too small an interval. At 8 times the diameter, which gives ⅔ the normal growth, growth is not higher than at 24 times the interval, or double the normal interval.

Even with 30 times the interval this table shows that there is a greater increase than with 7 times the interval.

We have in the above merely supposed that the isolated pine has the capacity to extend its roots to an interval 30 times its diameter.

Accordingly, in a wood where the trees were placed at an interval from each other 30 times their diameter, the growth of each tree would be equivalent to that of the free-standing, isolated tree.

Since for such free-standing trees it is possible to observe the extension of their roots as well as their growth in diameter we have an important means of testing and correcting the hypothesis we have made.

According to our method, the increase of diameter behaves like the effective interval. With an interval of 12 times the increase is $12 - 6 = 6$, at 30 times the interval it is $30 - 6 = 24$. The proportion between the two is therefore $6 : 24 = 1 : 4$.

If with a distance of 12 times the diameter the annual growth is $\frac{1}{6}$ of an inch, the free-standing pine would have an annual growth in diameter amounting to $4 \times \frac{1}{6} = \frac{2}{3}$ inch.

If, on the other hand, the capacity of the pine to extend its roots was limited to 24 times its diameter, the annual growth of the free-standing tree would be $3 \times \frac{1}{6} = \frac{1}{2}$ inch.

Some observations on the growth of trees:

1) On the edge of a local small pinewood sowed 38 years ago the diameters of 14 trees (measured 4 feet from the ground) amounted to 118 inches. That is 8.43 inches per tree. If one subtracts from the tree's age five years for its establishment: this gives per year a growth of $\frac{8.43}{33} = 0.225$ or approximately ¼ inch. In the interior of this wood a further 14 trees were selected, and these gave a sum of 67 inches. Per tree that is 4.8 inches, the annual growth being $\frac{4.8}{33} = 0.145$ or about $\frac{1}{7}$ inch.

2) Along an avenue 12 feet wide with trees at an interval of 16 feet from each other, 23 years after planting 10 birches have together a diameter of 87 inches, so that each tree is on average 8.7 inches thick. When planted the birches were probably about 1½ inches in diameter.

The growth in 23 years is therefore 7.2 inches, which is an annual figure of 0.313 or almost ⅓ of an inch.

3) In a small oak wood felled 25 years ago, where the oaks were perhaps between 80 and 100 years old, one thinner tree was left standing because of its strikingly modest growth; its diameter at the time I estimate to have been 7½ inches. Now, 25 years later, this tree, measured at four feet from the ground, has a diameter of 17½ inches. Its growth is therefore $^{10}\!/_{25} = \frac{2}{5}$ of an inch annually. This tree is in no respect free-standing, but only dominant, since it is now surrounded by 25-year-old oaks and chestnuts, which limit the extension of its roots. It is, however, notable that the height to which this tree has grown has not stayed in proportion to its increase in girth.

4) On some soil very favourable to trees a Canadian poplar planted 20 years ago now has a diameter of 21 inches. When planted it was perhaps 2 inches broad. The growth is therefore 19 inches in 20 years, 0.95 or almost one inch per year. The breadth of the avenue is only 10 feet, and the trees are placed 16 feet apart.

One poplar standing at the end of this avenue, isolated on two sides, now has a diameter of 23 inches, amounting to a growth of 1.05 inches per year.

These notes are insufficient to determine the growth of free-standing pines; but I present them all the same, since I hope that experienced foresters will thereby be prompted to exchange their own observations.

§ 20.
Calculation of total growth.

Before moving on to this calculation, we have to look back at the course taken by our study.

As with the line of arguments founding Part One of the *Isolated State*, we base our study on simple assumptions, dealing first of all with solely one of the factors of forestry, abstracting for the time being from all others, before later making them in turn the sole object of our investigations.

Thus we have in Sections One and Two assumed that the total growth of trees, however distributed in the wood, is 150 cubic feet per *Waldmorgen* per year; and from this we have established how ground rent and the most advantageous rotation period are affected by thinning more or less vigorously the timber stands and hence the capital stock that the timber represents.

From this we have learned what a great effect the greater or lesser clearance between trees has on ground rent and the most advantageous rotation period, even if *this clear interval had no influence on the tree's growth*.

Now that we have found a formula for the relationship between the interval of trees from each other and their growth we are now in a position to investigate the influence that stronger or weaker thinning has on growth, and hence on the timber yield of the entire wood.

Instead of taking our earlier hypothetical growth of 150 cubic feet we can, on the basis of the foregoing, calculate total growth for any interval of trees from each other:

1. For diameter δ the volume of the tree is $31\delta^3$.
2. If the interval of trees from each other is $y\,\delta$ and the minimum interval required to maintain the tree is $m\,\delta$, the growth in diameter is $\dfrac{y-m}{r-m} \times \alpha$ feet, and the growth of a woodland area of w square feet $= \left(\dfrac{31w\alpha}{r-m}\right)\left(\dfrac{y-m}{y^2}\right)$ cubic feet, where r and α can be considered as given by experience.

3. The maximum growth of timber occurs when $y = 2m$. It is known from practice that where the trees are placed such that $r = 12$ the increase in diameter $\alpha = \frac{1}{72}$ feet, then if one takes the above values for r and α and assumes that $m = \frac{1}{2}y$, this formula $\left(\dfrac{31w\alpha}{r-m}\right)\left(\dfrac{y-m}{y^2}\right)$ becomes $\left(\dfrac{31w}{72(12-\frac{1}{2}y)}\right)\left(\dfrac{y-\frac{1}{2}y}{y^2}\right)$

$$= \left(\frac{31w}{72(12-\frac{1}{2}y)}\right)\frac{1}{2y}$$

$$= \frac{31w}{72}\left(\frac{1}{24y-y^2}\right)$$

For what value of y does this function reach its maximum?

Taking the differential of this and setting it equal to 0 gives

$$\frac{31w}{72}(-24dy + 2ydy) = 0$$

Thus $2y = 24$
$$y = 12.$$

The growth of the remaining stock is

$$\frac{31w}{72}\left(\frac{1}{24y-y^2}\right) \quad \text{and if } y \text{ is } 12$$

$$\frac{31w}{72}\left(\frac{1}{288-144}\right) = \frac{31w}{72\times144} \quad \text{cubic feet.}$$

If $w = a$ *Waldmorgen* of 130 square rods $= 130 \times 256$ sq. ft., then the growth is $\frac{31 \times 130 \times 256}{72 \times 144} = \frac{31 \times 130 \times 2}{9 \times 9} =$ 99.5 cubic feet, which agrees with the table in the preceding section.

4. From the year the total growth is three times that of the increase in the remaining stock.

From the 6th to tenth year inclusive growth contributes $\frac{2}{3}$ of the later total growth. The growth over the first five years is joined to that in the sixth and reckoned as worth one year. The growth during the first five years does not therefore enter into the calculations.

The wood yielded from the thinning is from the 11th year twice the amount of the increase in the remaining stock, while from the 6th to 11th year it is equal to it.

Application of these propositions to concrete cases.

A. Timber stocks on clear felling.

Age at felling is	77 years,
subtracting the first	5 years
of growth gives	72 years.

The diameter of the tree increases annually by $\frac{1}{6}$ of an inch $= \frac{1}{72}$ feet. This gives the 77-year-old tree a diameter of $72 \times \frac{1}{72} = 1$ foot.

The volume of the tree is $31\delta^3$, hence here 31 cubic feet. If the trees are placed at 12 times their diameter from each other we have on a *Waldmorgen* of 130×256 sq. ft.

$$\frac{130 \times 256}{144} = \frac{130 \times 16}{9} = 231 \text{ of such trees.}$$

The timber stock from clear felling is therefore $231 \times 31 = 7,161$ cubic feet.

From the table in § 18 the timber stock from felling at any age can be calculated, taking due account of the first five years as above and multiplying the remaining years with the growth given in the table. So for example a tree felled at 77 years and placed at an interval from its neighbouring trees 12 times the diameter of its trunk and having a growth of 99½ cubic feet gives a timber stock per *Waldmorgen* of 72 × 99½ = 7,164 cubic feet.

In the first section I calculated according to these observations a growth of 100 cubic feet on 130 square rods. According to Nagel's propositions, where 12 times the interval between trees leads to an annual growth of $\frac{1}{6}$ inch, this gives an annual increase of timber of 99½ cubic feet on 130 square rods. It is remarkable that two such calculations, made on the basis of such different sources, should agree so nearly in their results.

In his *Hülfstafeln* p. 42 Cotta gives for a timber stock of 77-year-old pines on the best soil 14,860 cubic feet per Saxon acre. Converted to a Mecklenburg measure this is

14,860 × $\frac{36}{100}$ = 5,350 cubic feet per 100 square rods
or 6,955 cubic feet per *Waldmorgen* of 130 sq. rods.

Here, therefore, between the results of our calculation and the findings of an eminent forester, there is a difference of only 3%.

In the following calculations I take for the growth per *Waldmorgen* instead of 99½ cubic feet the round number 100.

B. Timber yield from thinning.

If the timber growth in the remaining stock is 100 cubic feet per *Waldmorgen*, then after the 11th year thinning yields annually 200 cubic feet, from the sixth to the tenth year inclusive only 100 cubic feet.

The total increase in a wood aged $x + 5$ years is:

1) from clear felling	$100x$ cubic feet
2) from thinning during the 6th to 10th years	500 cubic feet

3) from thinning from year 11

until clear felling $(x - 5)200$	$\underline{200x - 1,000}$
Total	$300x - 500$

An annual average of $\dfrac{300x - 500}{x + 5}$

For $x = 72$ that gives 274 cubic feet.

§ 21.
Ground rent and most advantageous rotation period for different degrees of interval between trees.

A. For an interval equal to 12 times the diameter.

As we have seen, the growth of the remaining timber stock is 100 cubic feet per *Waldmorgen*, and for the rotation period $x + 5$ years, according to § 3 the value of clear felling is $(3x^2 + 15x)$ sh.

From this there is

1) for the first five years from
   year 6 to 10 inclusive   150
2) from year 11 until felling $3x^2 + 15x - 150$

If the thinning wood had the same value per cubic foot as the felled timber, the product of thinning would have the following value:

1) Thinning in the 5 years from
   6th to 11th year give 100 cubic feet
   annually, whose value according to § 3 is   150.
2) Thinning from the 11th year until
   clear felling gives 200 cubic feet
   annually, whose value is
   $2(3x^2 + 15x - 150)$ or $\underline{6x^2 + 30x - 300}$
   together $6x^2 + 30x - 150.$

But since the thinning wood only has $\frac{2}{3}$ the value of the clear felled timber, the value of all thinnings is reduced to

$$\frac{2}{3}(6x^2 + 30x - 150)$$
$$= 4x^2 + 20x - 100.$$

Added to the dear felled timber $= 3x^2 + 15x$

Yields the sum of revenues $7x^2 + 35x - 100.$

Expenditure:

The value of all timber stocks is according to § 3

$$= x^3 + 9x^2 + 8x.$$

Interest at 4% on this is $0.04x^3 + 0.36x^2 + 0.32x$

The costs of administering
and supervising $x + 5$ *Waldmorgen*
at 8 sh. $8.00x +$ 40

Seeding costs of
130 square rods at 2 sh. 260

Sum of expenses is $0.04x^3 + 0.36x^2 + 8.32x + 300$

After subtracting these expenses from the revenue we have a ground rent of $-0.04x^3 + 6.64x^2 + 26.68 - 400.$

If we substitute a series of values for x we obtain

Rotation of	Ground Rent per *Waldmorgen*
15 years	32.7 sh.
25 years	98.8 sh.
35 years	151.3 sh.
45 years	194.0 sh.
55 years	227.9 sh.
65 years	253.3 sh.
75 years	270.4 sh.
85 years	279.4 sh.
90 years	280.8 sh.
95 years	280.3 sh.
105 years	273.0 sh.

Here the most advantageous rotation period is at the age of 90 years. Ground rent is then 280.8 sh.= 5 Th. 40.8 sh. per *Waldmorgen*.

B. For an interval 8 times the diameter.

For this interval the growth of the remaining stock is, according to § 19, 74.6 cubic feet per *Waldmorgen*, which we here round up to 75 cubic feet.

Hence the value of clear felling, thinning and the entire timber stock is ¾ of that which we have calculated in A for an interval of 12 times the diameter, and we accordingly obtain the following general formula:

Value of clear felling	$2.25x^2 + 11.25x$
Value of thinnings	$\underline{3.00x^2 + 15.00x - 75}$
Revenue	$5.25x^2 + 26.25x - 75$

Expenditure:
Interest on the value of the timber stock
$$= 0.03x^3 + 0.27x^2 + 0.24x$$

Costs of administration and supervision	$8.00x + \quad 40$
Seeding costs	$\underline{\hspace{3cm} 260}$
Sum of expenses	$0.03x^3 + 0.27x^2 + 8.24x + 300$

Subtracting this from the revenue
gives a ground rent of $-0.03x^3 + 4.98x^2 + 18.01x - 375$
    This formula gives

For a Rotation of $x + 5 =$	Ground Rent per *Waldmorgen*
55 years	167.7 sh.
65 years	187.0 sh.
75 years	200.0 sh.
85 years	206.8 sh.
92 years	207.7 sh.
95 years	207.5 sh.
105 years	202.2 sh.

Here a rotation period of 92 years proves to be the most advantageous, producing a ground rent of 207.7 sh. per *Waldmorgen*.

C. Interval between the trees is 7 times their diameter.

According to § 19 the growth of the remaining stock for this interval is 48.7 cubic feet. If we round this to 50 cubic feet we can find the calculation of ground rent already presented in § 18, for which the most advantageous rotation period is at age 93, with a ground rent of 135 sh. per *Waldmorgen*.

Results

Interval expressed in diameters	Growth of remaining stock per *Waldmorgen*	Rotation period in Years	Ground rent per *Waldmorgen* (sh.)
12	100 cu. ft.	90	280.8
8	75	92	207.7
7	50	93	135.0

It is apparent here that a reduction in the interval between trees extends the most advantageous rotation period, but not substantially, and that the ground rent diminishes rather more than the growth.

Where intervals differ it is, however, not the age of the tree, but its diameter that is the measure of the value of timber per cubic foot. Since the diameter is in proportion to the total growth of the woodland, for the interval of 8 times the diameter the ground rent falls to about half, and for an interval of 7 times to about ¼ of the rent that the land yields for an interval of 12 times the diameter.

Woodland rent when the trees are at
an interval of 12 times.

Revenue

1) from clear felling $\qquad$ $3x^2 + 15x$
2) from thinnings as
   above $2/3(6x^2 + 30x - 150)$ $\qquad$ $= 4x^2 + 20x - 100$

$\qquad\qquad$ Total $\qquad$ $7x^2 + 35x - 100$

Expenditure

Administration and supervision costs
for $x + 5$ *Waldmorgen* at 8 sh. $\qquad$ $8x + 40$
Seeding costs for 130 sq. rods at 2 sh. $\qquad$ $\underline{260}$

$\qquad\qquad$ Expenses $\qquad$ $8x + 300$

Subtracting these expenses there
remains a woodland rent for $x + 5$
Morgen of 130 sq. rods of $\qquad$ $7x^2 + 27x - 400$

This gives:

Rotation of	Woodland rent per *Waldmorgen* (sh.)
15 years	38
25 years	118
35 years	192
45 years	264
55 years	335
65 years	406
75 years	477
85 years	548
90 years	583
95 years	618
105 years	688

The woodland rent is made up of two elements: 1) ground rent and 2) the interest on the capital embodied in the timber stock. The following table shows the relation of these two elements to each other with differing rotation periods.

Rotation period	Ground rent (sh.) per *Waldmorgen*	Interest on capital embodied per *Waldmorgen*
15 years	32.7	5.3
25 years	98.8	19.2
35 years	151.3	40.7
45 years	194.0	70.0
55 years	227.9	107.1
65 years	253.3	152.7
75 years	270.4	206.6
85 years	279.4	268.6
90 years	280.8	302.2
95 years	280.3	337.7
105 years	273.0	415.0

While ground rent reaches its maximum with a rotation of 90 years, the woodland rent constantly increases. But this increase comes from the greater capital cost, and with a very high rotation period the woodland rent comes entirely from interest upon the timber stock.

§ 22.
What part of the timber reserve is at different ages removed in decennial thinnings?

With an interval between trees 12 times their diameter the total annual growth per *Waldmorgen* is from the 6th to the 10th years inclusive   200 cubic feet and from the 11th year onward   300 cubic feet. From the 6th year 100 cubic feet of this go annually into

the standing stock and the rest is removed in thinning.
1. At age 15 the stock comprises:

a) growth from the 6th to 11th year   1,000 cubic feet
b) growth from 11th to 15th year
   inclusive   1,500 cubic feet
                                        Stock   2,500

Standing stock   1,000
There remains 1,500 cubic feet for thinning, hence $\frac{3}{5}$ of
the standing stock.
2. Thinning at age 25 years
Ten years' growth contributes   3,000
                                        Stock   4,000

Of this to the standing stock there remains   2,000
Thinning gives 2,000 cubic feet, hence half of the
reserve.
3. Thinning at age 35 years:

Growth gives in 10 years   3,000
Reserve   5,000
Of this remaining stock   3,000
Thinnings give 2,000 cubic feet, hence $\frac{2}{5}$ of the
reserve.

Continuation of these calculations shows that the portion of the reserve that is affected by the thinning is as follows:

At the 4th thinning at age of 45 years: $\frac{1}{3}$
At the 5th thinning at age of 55 years: $\frac{2}{7}$
At the 6th thinning at age of 65 years: $\frac{1}{4}$
At the 7th thinning at age of 75 years: $\frac{2}{9}$
At the 8th thinning at age of 85 years: $\frac{1}{5}$
At the 9th thinning at age of 95 years: $\frac{2}{11}$

§ 23.
Deviation of the results of
our calculation from reality.

Our investigation is based on the following presuppositions:

1. that thinning is done continuously and proportioned to the growth;
2. that all trees are of the same health, vigour and strength;
3. that all trees stand in squares, have everywhere a normal interval and that until clear felling there are no open areas.

In reality we find quite considerable deviations from these conditions, all of which contribute to the diminution of the timber yield.

Among these deviations these are probably the most important:

1) if thinning is not done constantly, but periodically – as is everywhere actually the case – then trees can only enjoy the normal interval in the mid-part of the period, the growth being weaker in the years before and after this point than we have calculated. This reduction in yield, which can be calculated using the propositions in § 19, is not so very great if thinning is done on a ten-year cycle, but is very significant if it is only done every 20 years.

2) In wooded areas that have arisen from seeding one can seek through thinning that trees standing on a given area, e.g. a *Waldmorgen*, receive *as a whole* their normal space, but it is not possible to secure this normal space for every *single* tree. Trees will in some places be more crowded, in others more dispersed, than the

normal distribution that is required for growth to be its maximum.

If, for example, in one *Waldmorgen* the places where the trees are at an interval of ten times their diameter from each other make up half the *Waldmorgen*, while on the other half the trees are spaced at 14 times the diameter of their trunks, we can calculate their growth as in § 19:

$$\text{10 times distant } \frac{95.5}{2} = 47.75 \text{ cu.ft.}$$

$$\text{14 times distant } \frac{97.5}{2} = 48.75 \text{ cu.ft.}$$

The growth on one *Waldmorgen* is 96.5 cu. ft., whereas with a normal interval
for all trees it would be   99.5 cu. ft.

3) Accidental damage. Woods are subject to windthrow, fire, plagues of caterpillars, damage from larger animals or from hunting, etc.

If there were an insurance company through which one could insure against such incidents one would have to deduct the premium paid from the revenue. For lack of such a company the owner of the woodland has to take into account such losses by estimating their level on the basis of experience and deducting it from the revenue.

4) Theft. Theft of wood can never be completely avoided, and in this way a part of the estimated income is lost to the owner of woodland.

The timber yield that foresters themselves provide from what they find in the wood has already had such losses deducted from it, while they still have to be deducted from the yield that we calculate for an entirely regular stand of timber. This is one explanation of why our calculated yields are far above the figures reported by foresters. This difference does, however, also stem from the fact that foresters are not working with woodland in which there are stands of timber that have been regularly thinned since they were saplings, for which reason their calculations of yield based on such practical conditions cannot be applied to such stands.

The deviations represented by 1) and 2) above can be determined fairly exactly with the calculation made in § 19. The deviations in 3) and 4) by contrast depend entirely on the position of the trees, other local conditions and the character of the region's inhabitants, and have to be assessed for every specific location.

But that is the great advantage we have from theoretical investigation: we gain a solid basis with which all other conditions can be compared, and hence related back to a uniform viewpoint.

For the proportion that I assume here the diminution of timber yield involves a deviation from ideal conditions and the estimated revenue of $\frac{1}{3}$. At the same time interest on the capital embodied in the timber stands is reduced by $\frac{1}{3}$.

Our calculation is based on the implicit assumption that all kinds of wood – from wood strips and bean poles to entire beams of timber – will always find a market and can be sold every year. But such an extensive and assured market is only to be found close to larger towns and cities, and near navigable canals and rivers.

If such a favourable situation is lacking the sale of building and other timber materials depends upon chance – whether buildings are being put up within a few miles more or less, whether fencing is being done, etc. The regular sale of wood is in such circumstances quite uncertain.

If thinning is regularly completed – which is necessary if the calculated timber yield is to be achieved – then in many years some of the building and other useable timber will have to be chopped up and corded as firewood to be sold at much lower prices; or, alternatively, timber stores have to be built which in turn involve interest and storage costs that have to be accounted for. The losses arising from such circumstances have to be estimated separately for each location.

For the location that I presuppose I assume the reduction in revenue arising from the reduction in timber value to be ¼ of the calculated revenue in § 21. Consequently, from the revenue given in § 21 should be deducted:

1. for the reduction in timber yield   33 ⅓%
   one-third or leaving a remainder of   66⅔%
2. for the decrease in timber value
   one-quarter or 25%, this is for 66 2/3%   16⅔%,
   hence remaining   50%

Hence 50 remains from every 100 in revenue calculated in the Table given in § 21.

In all comparisons with reality I thus take account of just half of the revenue calculated in § 21.

§ 24.
Comparison of the ground rent of woodland with that of arable land.

According to § 21, where the trees are spaced normally the revenue per *Waldmorgen* is:

1) From clear felling	$(3x^2 + 15x)$ sh.
2) From thinnings	$4x^2 + 20x - 100$
Sum	$7x^2 + 35x - 100$

From this there is deducted:
The interest on
the capital stock $\underline{\qquad 0.04x^3 + 0.36x^2 + \quad 0.32x \qquad}$
Leaving revenue of $-0.04x^3 + 6.64x^2 + 34.68x - 100$
For a rotation of $x + 5 = 90$ years this gives an annual revenue of 290.8 sh. (If one deducts from this 10 sh. for administration and seeding costs, then this gives a ground rent of 280.8 sh. as in § 21).

On account of the conditions outlined in the preceding paragraphs revenues are however one half of the estimated amounts, and so make up $\dfrac{290.8}{2} = 145.4$ sh., or 139.8 Th. from 6,000 square rods.

Deducting from this the administration and seeding costs per *Waldmorgen* of 10 sh. or 9.6 Th. per 6,000 square rods; this gives a ground rent per *Waldmorgen* of 135.4 sh.

or per 100 square rods   104.2 sh.
or per 6,000 square rods 6,252   sh.
or   130.2 Th.

The rent of arable land.

For land that, following a fallow period,
yields 10 corns in rye

(10 Berlin bushels per 100 square rods).

If this field is on average 210 rods from the farmyard, then (according to § 11 of Part One of the *Isolated State*) once the interest on capital (represented by buildings and other objects separable from the land) is subtracted from the net revenue, a field of 70,000 square rods gives a ground rent of 954 Th.

For a field of 6,000 square rods (about 50 Magdeburg *Morgen*) this amounts to 81.8 Th.

By contrast, we have found that the ground rent yielded by an area of 6,000 square rods devoted to forestry is 130.2 Th.

We arrive therefore at the surprising and very remarkable result that good, well-fertilized soil sown with rye and yielding 10 corns gives, if sown with pines, a rent 60% higher than if used for grain.

This relationship is even more striking if we consider less fertile land.

Assuming that for different types of land the timber yield is in direct proportion to the grain yield on such land, then we have the following results.

1) On land yielding 9 corns.

The timber yield and hence the revenue from timber is here $\frac{1}{10}$ less than on land yielding 10 Corns and yields therefore $\frac{9}{10} \times 139.8 =$ 125.8 Th. from 6,000 square rods.

Deducting for costs of
seeding and administration   9.6 Th.
leaves a ground rent on
6,000 square rods of   116.2 Th.
The ground rent is (according to *Isolated State* Part One § 11) 65.1 Th.

2) On land yielding 8 corns.

Revenue is $\frac{8}{10} \times 139.8$ 111.8 Th.
Deducting for Seeding etc.   9.6 Th.

Ground rent   102.2 Th.
Ground rent from arable use of 6,000 square rods yields 48.5 Th.

3) On land yielding 7 corns.

Revenue is $\frac{7}{10} \times 139.8$ 97.8 Th.
Deducting for Seeding etc.   9.6 Th.

Ground rent   88.2 Th.

Ground rent from arable use of 6,000 square rods yields 31.9 Th.

4) On land yielding 6 corns.

Revenue is $\frac{6}{10} \times 139.8$ 83.8 Th.

Deducting for Seeding etc.   9.6 Th.

Ground rent   74.2 Th.

Ground rent from arable use of 6,000 square rods yields 14.3 Th.

Woodland rent.

This consists of ground rent and the interest on the capital embodied in the timber stock, and is calculated by subtracting the costs of seeding and administration from the gross revenue from timber.

The revenue is:

1) from clear felling (§ 21)	$3x^2 + 15x$
2) from thinnings	$4x^2 + 20x - 100$
Total	$7x^2 + 35x - 100$

For a 90-year rotation in which $x = 85$, this gives per *Waldmorgen* 593 sh. per 6,000 square rods = 570.2 Th.

Given the deviations noted in the preceding paragraphs only one half of this
is to be counted   = 285.1 Th.

Deducing for seeding and
administration costs   9.6 Th.

results in a Woodland Rent of   275.5 Th.

Land yielding 9 Corns gives
a revenue of $\frac{9}{10} \times 285.5$ = 256.6 Th.

Subtracting the noted
costs of   9.6 Th.

gives a woodland rent of   246 Th.

Extending this calculation gives:

Land of 8 corns yield has
a woodland rent of   218.5 Th.
Land of 7 corns yield has
a woodland rent of   190.0 Th.
Land of 6 corns yield has
a woodland rent of   161.5 Th.

Summary.

An area of 6,000 square rods yields, if forested with pines on a 90-year rotation or used as arable:

Yield of land in corns	Ground rent as woodland (Th.)	Ground rent as arable land (Th.)	Relation between the two	Woodland rent (Th.)
10	130.2	81.8	160 : 100[1]	275.5
9	116.2	65.1	179 : 100	247.0
8	102.2	48.5	211 : 100	218.5
7	88.2	31.9	276 : 100	190.0
6	74.2	15.3	485 : 100	161.5

On less fertile lands giving 5, 4 and 3 corns the yield of forestry in comparison with agriculture is even more considerable. But perhaps on such soils during the first rotation no good-quality building timber can be grown, and this leads to a quite different calculation.

The results of our calculation contrasts most starkly with the local prevailing view that pine silviculture can be pursued with advantage only on soils of no arable value, and that pines should be removed from all better land.

---

1. In this comparison the yield of arable land is always equivalent to 100.

For the high level of deductions made in § 23 for the deviations of reality from our ideal presuppositions it is in my view possible in most cases to estimate the deductions rather lower, and the yields higher, than there calculated. It would be interesting to explore the origin of this generally accepted view that I regard to be entirely in error wherever building timber has the usual price assumed.

It probably originates in defective treatment of woodland, specifically in faulty thinning methods, in which not merely is timber growth little diminished, but in which the temporary uses of wood is largely wasted – and so, based on such experience, the generally accepted view forms that is passed on to later generations without any serious examination.

Our study also shows the absence of any basis for the concern that building timber must rise in price with the gradual disappearance of ancient woodland. On the contrary, when knowledge of rational forest economy has become more widely known and is everywhere put to good use the price of 4 sh. $N\frac{2}{3}$ for a cubic foot of building timber will not last, and will have to fall.

§ 25.
Applications.

A. In general relation to the Mecklenburg context.

If a town meets its need for timber in part from an area one mile away, and in part from an area five miles away, paying the same for both, the value of the timber from the closer wood is higher than that from the more distant wood by a sum equivalent to the transport costs over four miles. This difference is very considerable, and – as proved in the first part of the *Isolated State* – is much greater that the difference in value of grain from a farm one mile away from the town, and that from one five miles away.

Variation in land and ground rent is closely related to this variation in the value of products.

If in the area five miles removed from the town the ground rent from forestry exceeds that from arable cultivation, this excess will be even greater closer to the town.

Rationally therefore, the land closer to the town should be more heavily wooded than that at a greater interval.

If we consider the larger cities in Mecklenburg, such as Rostock, Schwerin and Wismar, we can see that in actuality we find the exact opposite. The closer we come to these cities, the sparser becomes any woodland, and along the country road from Laage to Rostock one sees, with few exceptions, only bare, treeless fields, while on the road from Teterow to Laage, five or six miles distant from Rostock, where timber has a far lesser value, we are surrounded by woodland. The larger cities therefore appear to lay waste woodland, whereas they should naturally stimulate the growth of silviculture.

What a contradiction we have here between theory and practice!

This results in Rostock, before the construction of the country road, a cord of beechwood was worth from 14 to 16 Th. $N\frac{2}{3}$. Much worse than this is that along the Baltic coast of Mecklenburg the need for pine or fir building timber is not met by domestic production, but instead has to be imported from Sweden, Norway and Finland.

A respected merchant in Rostock who had an important timber business let me have these details on timber imports into Mecklenburg:

That he estimated the annual value of timber imported into Rostock from northern ports at

50,000 Thaler N$\frac{2}{3}$, and that imports into Wismar would be very similar. In addition, a considerable quantity of timber boards was imported into northern Meckenburg through Demmin.

Since there is sufficient land in northern Mecklenburg more suited to the cultivation of pines than grain, then if landowners were to follow their own interest and recognize the value of rational silviculture they would in future save the payment of at least 100,000 Thalers which at present flows abroad in payment for building timber. That is not to join with the petty standpoint that all saving in the export of precious metals represents an increase of national income; for the opposite can happen, as the following example shows.

Mecklenburg perhaps pays just as much overseas for strong timber imported from Angeln and Jutland as it does for imported building timber. This money could obviously be kept at home and employed in raising cattle. But if calves are to be bred here this will mean fewer milk and pasture herds, and if this reduces the export of butter and fat cattle by 120,000 Thalers, saving the expenditure abroad of 100,000 Thalers is not only no gain, but a loss of 20,000 Thalers.

Things are very different for the money that is paid abroad for timber; for according to the calculations made in the preceding paragraphs land that yields six corns pays in rent through forestry four times that which it pays under arable cultivation, producing timber for which at present 100,000 Thalers flows overseas, sacrificing at the most 25,000 Thalers.

The domestic production of timber would therefore raise national income by at least 75,000 Thalers.

However significant this increase might be, there is in the south of Mecklenburg a much richer source for the increase of national income.

In southern Mecklenburg a very sandy soil is predominant, yielding a very low rent from agriculture; and the opening up of the Elbe and Havel to navigation has made possible the sale of wood to Berlin and Hamburg, two markets where there is no need to fear over-supply. It is in fact a pleasing thought that the extension and improvement of Mecklenburg's forest culture represents such prospects for the increase of national income.

B. In relation to the special conditions of the Tellow estate.

In § 21 we found that the relationship in ground rent between a 90- and a 35-year rotation was of the order of $280.8 : 151 = 100 : 54$. A lengthy rotation is therefore far more advantageous than a short one. Nonetheless, I have introduced a 35-year rotation on the Tellow estate. This is not a contradiction between knowledge and practice, ignoring the views expressed above; and is explained by local conditions around the Tellow estate which I will now briefly outline.

Pine woodlands are scattered in different areas and together take up 13,000 square rods (110 Magdeburg Morgen).

The land in these woodlands is very variable, consisting partly of sand, partly of good arable land with barren areas and clay hillocks overlaying marl.

At the beginning the pines grew very quickly on all types of soil with the exception of the barren areas. At the age of 30 the trees growing on clay, which had gradually become covered with wild raspberries (*Rubus idaeus*), became sickly and gradually died. In this way quite unpleasant gaps opened up which became gradually more extensive, and so I was, against my will, compelled to introduce a short rotation.

Continuation.

Continuation of this work will have to work through the following topics:

Section four.

§ 26. Assessment of the Herr Cotta's view that private owners of woodland always have an interest in devastating their woodland.

§ 27. Comparison of our calculations with the approaches and data of various foresters.
A. in relation to the timber yield of clear felling
B. in relation to the interval between trees
C. in relation to the yield from thinning out.
Comparison with the thinning yield of Herr Hartig, if possible to ascertain why foresters assume that the yield from thinning is so unnaturally small.

§ 28. If for all circumstances only $\frac{1}{3}$ of the total growth can remain in the standing timber, what happens to the remaining $\frac{2}{3}$ if thinning is not carried out everywhere?

Section five.

§ 29. How is one to estimate the value of land intended for afforestation that has not yet been planted?
  A. Herr Nagel's method and communication of his calculation.
  But this calculation involves a contradiction with the principle that rotation should not be extended beyond the point where the highest ground rent is to be found.

Nagel assumes a 120-year rotation and estimates the value of 1000 square rods of land at 300.w3 Thalers, from which the seeding costs have to be subtracted.

In his view this 1000 square rods would have a value with a 30-year rotation of 4 × 212.6 = 850.4 Th. and with a 60-year rotation 2 × 307.1 = 614.2 Th., from which the seeding costs have to be subtracted.

With a 60-year rotation therefore the land would have approximately double the value as with a 120-year rotation.

§ 30. Calculation of the value of land according to our propositions.

§ 31. For which rotation do we here get the highest land value?[1]

§ 34. Comparison of the established land value with the ground rent of existing, regularly felled woods.

Section six.

Transition to the isolated state.

§ 35. Beyond the ring of forestry, timber is not bought and sold. Each estate produces for its own needs

---

1. Editorial addition: §§ 32 and 33 do not exist in Thünen's manuscript.

only. The task here is to establish for which thinning method and which rotation period timber is produced with the least cost.

Here the ground rent that could be achieved if the land were devoted to grain production becomes an element in the production price of timber.

The rent of land becomes ever smaller, the greater the distance from the town, and the question arises: what influence does a greater or lesser rent of land have on forestry?

§ 36. Propositions for the determination of the price of timber of varying size. The land must bear the same land rent, and can with a short rotation period produce weak timber, and with a longer rotation strong timber. In this respect the question is easy to answer. But with a longer rotation period weaker wood is a by-product of thinning.

With this the production costs cease to be a measure of prices, and the doctrine of the price of goods loses relevance. But this price cannot be purely arbitrary; and so what motivating elements determine its price? If this question is thought to be a general one, and not addressed to particular localities, it is extremely difficult to answer. Perhaps some support might be found in the comparison of estimates of the value of unsown land with the ground rent afforded by a well-tended forest.

§ 37. What influence does the rise and fall of urban timber prices have on the management of forests?

Production costs for beechwood as firewood of kinds of soil giving different land rent.

Since beech can only be grown on better land which, on *level ground*, produces a considerable land rent, the production price of beech will be so high that as ancient woodland disappears it will vanish from our woods and in the future be found only as an ornamental garden tree. It follows from this how contrary to economic doctrine it is to establish a servitude regarding the delivery of beechwood in perpetuity, and how thoughtless the Mecklenburg Chamber of Deputies was when it assumed responsibility for the delivery of a large quantity of beechwood in perpetuity.

The production costs of beechwood look quite different in mountainous regions, where a good clay soil, because of the steepness of the slopes makes them unsuited for agriculture and suited only for pasture, and as such produce only a very small rent.

Section seven.

§ 38. According to our estimation, the highest ground rent occurs with a 90-year rotation. 90-year-old timber does not make strong beams and no sawblocks. If we consider what prevents a longer rotation, the cause can be found in the strong growth of the interest on the capital stock.

The capital stock is however not an invariable magnitude, since we can significantly reduce it through spacing trees further from each other. But such spreading of the timber also reduces overall growth. But growth can be calculated for every interval between one tree and another as in § 19, and this enables us to include in our calculations both the reduction of interest and that of growth.

Those pines intended for use as building timber requires, if they are to achieve the form they require, need to be grown in closed stands. When they reach the necessary stem length and are growing straight, and if it is only a case of increasing their diameter, they can withstand any degree of separation without losing anything of their utility.

At 90 years trees have certainly reached the required shape and length.

If one assumes that after the 90th year each tree is given 24 times the interval to the next tree, instead of 12, then the capital stock declines by one quarter – not of the entire wood, but of the marginal compartment. With an interval of 24 times the diameter growth declines, according to § 19, from 99.5 to 74.6, that is, from 1 to ¾.

What influence does this alteration in the relative interval have upon ground rent?

According to § 21, a 90-year rotation gives a ground rent of 280.8 sh. per *Waldmorgen*. It can be asked whether the felling of trees more than 90 years old could make an equal or higher ground rent if the trees were given an interval of 24 times.

This question is answered if we subtract from the value of the growth of the marginal compartment the interest on the capital in *this* compartment.

With a 12 times interval felling at $x + 5$ years has a timber volume of 100 × cubic feet.

The value of 100 cubic feet = $3(x + 5)$.

The value of the felling is therefore $3x^2 + 15x$.

If this compartment is not felled, the value of the stock in the following year is:

Timber stock of x + 6th felling	$(x + 1)100$ cu. ft.
Value per 100 cu. ft.	$3(x + 6)$
Value of felling	$(x + 1)(x + 6)$
	$= 3x^2 + 21x + 18$

Subtracting the value of
the felled timber in the
preceding year as $3x^2 + 15x$

gives the one-year growth of
the remaining stock as $6x + 18$

According to § 21. p. 96 the capital stock is
$$x^3 + 9x^2 + 8x$$

If the rotation is extended by one year, x becomes $x + 1$. Hence:

$$
\begin{aligned}
x^3 &= x^3 + 3x^2 + 3x + 1 \\
9x^2 &= 9x^2 + 18x + 9 \\
8x &= 8x + 8
\end{aligned}
$$

Summing to: $\quad x^3 + 12x^2 + 29x + 18$

subtracting the earlier capital $\;x^3 + 9x^2 + 8x$

Gives the increase of capital: $\quad 3x^2 + 21x + 18$

We have included in the account for the $x + 6$th year only the timber value of the $x + 5$th year which is $\quad 3x^2 + 15x$

Which gives a difference of $\quad 6x + 18$

Which is precisely the increase on value of the previous year.

Revenue from clear felling and thinnings are together $7x^2 + 35x - 100$. If we substitute $x + 1$ for x in this, then

$$
\begin{aligned}
7x^2 &= 7x^2 + 14x + 7 \\
35x &= 35x + 35 \\
-100 &= - 100
\end{aligned}
$$

Summing to $\quad 7x^2 + 49x + 42 - 100$

Subtracting from the revenue for x years $\quad 7x^2 + 35x - 100$

Gives an increase in revenue of $\quad 14x + 42$

hence 4 times more than calculated above.

For $x = 85$ the surplus revenue is

$$
\begin{aligned}
14x &= 1{,}190 \\
+42 &= \phantom{1{,}1}42
\end{aligned}
$$

In total $\quad 1{,}232$

The additional contribution of capital is:

$$
\begin{aligned}
3x^2 &= 21{,}675 \\
21x &= 1{,}785 \\
+18 &= \phantom{21{,}67}18
\end{aligned}
$$

Which totals to $\quad 23{,}478$

The interest on this at 4% gives   939

                        Remains   293

Subtracting from the administration

costs of   8

leaving a ground rent of   285.

The surplus in the increased value of the clear felling with respect to the interest on the capital embodied in the felled timber must be as high as the ground rent which the rotation for the average of all fellings gives. Then if the surplus from clear felling were greater, to be consistent the rotation would have to be extended, if it were smaller, then the rotation would have to be reduced.

This leads to the question: for what rotation period is the surplus that clear felling gives equivalent to the average of the ground rent that the complete rotation produces?

The revenue is $14x + 42$ and subtracting 8 sh. for administration costs $= 14x + 34$.

The value of the $x + 5$th felling is $3x^2 + 15x$

Of this, increase in value during

one year is $6x + 18$

        The value of the $x + 6$th felling $3x^2 + 21x + 18$

Interest of $0.12x^2 + 0.84x + 0.72$

Subtracting this from the revenue of $14x + 34$

gives a yield from

the $x + 6$th felling of $-0.12x^2 + 13.16x + 33.28$

For what value of x does this surplus reach the medium ground rent of 280.8?

Putting them together gives:

$$-0.12x^2 + 13.16x + 33.28 = 280.8$$
$$-12x^2 + 1{,}316x + 3{,}328 = 28{,}080$$
$$- 3{,}328 = -3{,}328$$
$$-12x^2 + 1316x = 24{,}652$$
$$x^2 - 109.6x = -2{,}054$$

$$+ 3{,}003 = +3{,}003$$
$$x - 54.8 = \sqrt{949} = 30.8$$
$$x = 85.6$$

and the rotation $x + 5 = 90.6$ years.

The increase in value of the timber stock
is in one year $\qquad\qquad\qquad\qquad\qquad\qquad 6x + 18.$

The yield from thinning timber contributes
in mass double the increase in the clear felling,
by value however $2 \times \frac{2}{3} = 1\frac{1}{3}$ of the same,
hence $1\frac{1}{3}(6x + 18).$ $\qquad\qquad\qquad = \underline{8x + 24}$

$\qquad\qquad\qquad\qquad\qquad$ Revenue $\qquad = 14x + 42$

Deducting administration costs $\qquad\qquad\qquad \underline{8}$

$\qquad$ Leaves $\qquad\qquad\qquad\qquad\qquad\qquad 14x + 34$

The timber value of the $x + $ 5th felling is $3x^2 + 15x$. The
one-year growth of $6x + 18$ gives the timber value of the
$x + $ 6th felling at $3x^2 + 21x + 18.$

Deducting interest at 4% $\quad = \underline{-0.12x^2 + 0.84x + 0.72}$

leaves a surplus of $\qquad\qquad -0.12x^2 + 13.16x + 33.28.$

For $x = 85$ this becomes

$$-0.12x^2 = \qquad\qquad\quad -867$$
$$+13.16x = 1{,}118.60$$
$$+33.28 = \underline{\qquad 33.28}$$
$$= \underline{\qquad\quad 1{,}151.88}$$

leaves a ground rent of $\qquad\qquad\qquad 284.88$

How do revenue, capital stock and ground rent alter in
after the 90th year the wood is opened up in such a way
that the trees are spaced at an interval from each other of
24 times their girth?

Growth and also revenue from the one-year growth fall
according to § 19 in the proportion 99.5:75.6 = 4:3.

Annual income therefore becomes

$$\tfrac{3}{4}(14x + 42) = 10.5x + 31.5.$$

Taking into account administration costs of $\qquad\underline{8}$

leaves a revenue of $\qquad\qquad\qquad 10.5x + 23.5.$

If the trees which now have an interval of 12 times are given an interval of 24 times, only one quarter of the trees can remain and so the capital stock falls to one quarter, thus:

$$\tfrac{1}{4}(3x^2 + 21x + 18) = 0.75x^2 + 5.25x + 4.5$$

Interest contributes	$0.03x^2 + 0.21x + 0.18$
Subtracting these	
from the revenue of	$10.5\ x + 23.50$
gives a ground rent of	$-0.03x^2 + 10.29x + 23.32$

For $x = 85$ this becomes

$$
\begin{aligned}
-0.08x^2 &= &-216.75 \\
+10.29x &= &+874.65 \\
+23.32 &= &\underline{23.32} \\
&= &897.97
\end{aligned}
$$

leaves a ground rent of   681.22

We arrive at the very striking result that by spacing the trees out the ground rent increases from 285 to 681, more than double.

If the rotation period is 120, x is 115.
Hence

$$
\begin{aligned}
-0.08x^2 &= \tfrac{3}{100} \times 13{,}225 = &-396.75 \\
10.29x &= &+1{,}183.35 \\
+23.32x &= &\underline{23.32} \\
&= &1{,}206.67
\end{aligned}
$$

ground rent of 809.92

For $x = 135$ the rotation period is 140 years and

$$
\begin{aligned}
10.29x &= 1{,}389.15 \\
+23.32 &= \underline{23.32} \\
&= 1{,}412.47 \\
-0.03x^2 &= \underline{-546.75}
\end{aligned}
$$

ground rent   865.72

Ground rent is here constantly increasing and it can be asked at what value of x this finds its maximum. Setting the differential of the equation $-0.03x^2 + 10.29x + 23.32 = 0$, gives

$$-0.06x\mathrm{d}x + 10.29\mathrm{d}x = 0$$
$$0.06x \quad = 10.29$$
$$x = 171.5$$

and the most advantageous rotation period $= 176.5$ years.

According to this the 176-year rotation is more advantageous than the 140-year rotation. Since, however, there is very little demand for the strong wood that a 140-rotation period gives and its value per cubic foot no longer increases with increased girth; so according to this calculation the maximum of the ground rent is achieved with a rotation period of 140 years.

This calculation abstracts from the following related conditions:

1. if the interval between trees remains the same it is probable that after the 90th year the growth of individual trunks and of the entire wood gradually slows;

2. by contrast, by doubling the interval the annual growth of the individual trunks increase by more than 1/6 inches and with this the value of the wood increases per cubic foot to a greater degree than we have calculated;

3. with the greater amount of space more of the tree's growth goes into the branches and crown, with a lesser amount going into the trunk.

These varied conditions have to be taken into account in working through these propositions more carefully and completely – work which I have to leave to others.

Here, so that the calculation does not become involved, I have taken the same interval of 24 times for the 91-year and the 140-year rotation.

There is, however, no doubt that such thinning of the wood should not be done all at once, but gradually; and that for each age of tree, or rather, for each size of tree, a different interval must be applied if the maximum of ground rent is to be achieved in every year. The task is to find for every size of tree the interval for which the annual yield, having subtracted interest on capital stock, is the highest.

Previously we took the highest growth in value as the objective. But this objective changes here, in that we are now seeking the greatest surplus of the increase in value with respect to the interest on capital stock, hence the highest ground rent – not for the entire rotation, but for every single year.

The formula developed in § 19 regarding the relation between interval and the growth of trees puts us in a position where we can translate this task into calculation.

For the 12 times interval the growth of the remaining stock is 100 cubic feet per *Waldmorgen* and the annual increase in value including thinning is $14x + 42$.

For interval y the growth of the remaining stock of timber according to § 19 is $2,388\left(\dfrac{y-6}{y^2}\right)$ cu. ft.

Yield and value of timber growth is for both intervals 12 and y : $100 : 2388\left(\dfrac{y-6}{y^2}\right) = 1 : 23.88\left(\dfrac{y-6}{y^2}\right)$. For the interval 12 however the value is $14x + 34$. This gives for interval y the value for one year's timber growth

$$= 23.88\left(\frac{y-6}{y^2}\right)\left(14x + \overset{42}{34}\right).$$

(Revenue is $14x + 42$ and one cannot, as here, subtract 8 sh. administration costs in that these are not proportional to the timber yield, but constant.)

$$= \left(334x + 812 \overset{1,002}{} \right) \left(\frac{y - 6}{y^2} \right)$$

$$= \left(334x + 812 \overset{1,002}{} \right) y - 2,004x - 4,872 \overset{6,012}{}.$$

The timber value of the x + 5th felling is with
   a 12 times interval $3x^2 + 15x$
Adding the one year growth of $\underline{6x + 18}$
Gives a value for the x + 6th felling of $3x^2 + 21x + 18$
Interest contributes $0.12x^2 + 0.84x + 0.72$.
On the area $w\delta^2$ of a wood there can stand, with interval

of 12δ, $\dfrac{w\delta^2}{144\delta^2} = \dfrac{w}{144}$ trees. With an interval of $y\delta$ there

stand on this area $\dfrac{w\delta^2}{y^2\delta^2} = \dfrac{w}{y^2}$ trees of the same size as

with the 12 times interval.

The capital stock and the interest upon it are therefore in proportion to both intervals

$$\frac{w}{144} : \frac{w}{y^2} = \frac{1}{144} : \frac{1}{y^2} = 1 : \frac{144}{y^2}.$$

   Let us now take the year preceding the felling, rather than the year following it as the basis for our investigation.
   In the x + 5th year the value of the timber is per cubic feet $3(x + 5) = 3x + 15$
   The timber stock $= 100 \times$ cubic feet
   The value of the stock is therefore $3x^2 + 15x$
   In the previous year the value is per 100 cubic feet

$$= 3(x + 4) = 3x + 12$$

The stock is $= (x - 1)$ 100 cubic feet

The value of the stock is therefore $(x - 1)(3x + 12)$

$= 3x^2 + 12x + -(3x + 12) =$ $\quad 3x^2 + 9x - 12$

In the $x + 54$th year the value is $\quad \underline{3x^2 + 15x}$

The growth in value is therefore $\quad 6x + 12$

The yield from thinning in the final year of rotation is 200 cubic feet, or twice the timber transferred to the remaining stock. With an equal value per 100 cubic feet the yield from thinning would be $2(6x + 12) = 12x + 24$.

Since however the timber from thinning only has $\frac{2}{3}$ the value of the remaining timber, the yield from thinning is reduced by $\frac{2}{3}(12x + 24) = \qquad\qquad 8x + 16$.

Adding to this the value of

the remaining timber of $\qquad\qquad\qquad \underline{6x + 12}$

gives an annual yield of $\qquad\qquad\qquad 14x + 28$.

From this we have to subtract the interest from the value of the stock in the year it is felled. Since it is however doubtful whether interest can also be subtracted from the value of the growth in thinning timber we have to estimate the interest as in § 18.

According to § 18 the value of all timber stocks from the first up to the $x + 5$th year $= x^3 + 9x^2 + 8x$.

For the $x + 4$th year x becomes $x - 1$ and the expression alters as follows:

$$x^3 \text{ becomes } (x - 1)^3 = x^3 - 3x^2 + 3x - 1$$
$$9x^2 \text{ becomes } 9(x - 1)^2 = 9x^2 - 18x + 9$$
$$8x \text{ becomes } 8(x - 1) = \underline{8x - 8}$$
$$\phantom{8x \text{ becomes } 8(x - 1) =} x^3 + 6x^2 - 7x$$

If this value is subtracted from the

value of the following year it

becomes $\qquad\qquad\qquad\qquad x^3 + 9x^2 + 8x$

The increase of capital is $\qquad\qquad 3x^2 + 15x$

Interest contributes $\qquad\qquad\quad 0.12x^2 + 0.6x$

Subtracting this from the revenue = $\underline{14\quad x + 28}$

Gives a surplus of $\qquad -0.12x^2 + 13.4x + 28$

For $x = 85$ this becomes

$$-0.12x^2 = \quad -867$$
$$13.4\ x\ = 1139$$
$$+28\quad = \underline{\quad 28}$$
$$\underline{1167}$$

yields surplus $\qquad\qquad 300$

Subtracting 8 sh. for administration costs, this gives a ground rent of 292.

The average ground rent is however only 280.8.

The ground rent for 90 fellings

is accordingly 90 × 280.8 = $\qquad$ 25,272

The 90th year alone gives $\qquad\qquad \underline{\quad 292}$

Which leaves for the 89th year $\qquad$ 24,980

Which for 1 year is $\qquad\qquad$ 280.67.

The small difference perhaps stems from the fact that the most advantageous rotation period is not exactly 90 years, but slightly higher.

For the 12 times interval the growth is 100 cubic feet.

For y times the interval according to § 19 the growth in the remaining stock is $2388\left(\dfrac{y-6}{y^2}\right)$ cubic feet.

The proportion between both intervals is therefore

$$100 : 2388\left(\frac{y-6}{y^2}\right) = 1 : 23.88\left(\frac{y-6}{y^2}\right)$$

For the interval 12 we have found a value for growth as follows: $14x + 28$.

For interval y therefore the value of growth is

$$= 23.88\left(\frac{y-6}{y^2}\right)(14x + 28)$$

$$= (334x + 668)\left(\frac{y - 6}{y^2} \right)$$

$$= \frac{(334x + 668)y - 2004x - 4008}{y^2}.$$

Relation between the capital stock of the x + 5th felling at intervals of 12 times and y times.

The girth of the trees is δ, so on the wooded area $w\delta^2$ there stand:

A. with the 12 times interval $\dfrac{w\delta^2}{144\delta^2} = \dfrac{w}{144}$ trees

B. with the y times interval $\dfrac{w\delta^2}{y^2\delta^2} = \dfrac{w}{y^2}$ trees

Since it is here assumed that trees have the same diameter, the stock and the number of trees are in direct proportion. With an interval of 12 and of y the stock is therefore $\dfrac{w}{144} : \dfrac{w}{y^2} = 1 : \dfrac{144}{y^2}$.

For the 12 times interval the contribution of interest on the capital stock is $0.12x^2 + 0.6x$.

For the y times interval this gives in interest

$$\left(0.12x^2 + 0.6x\right)\frac{144}{y^2} = \frac{17.3x^2 + 86.5x}{y^2}.$$

Subtracting this from the revenue of

$$\frac{(334x + 668)y - 2,004x - 4,008}{y^2}$$

gives $\dfrac{(334x+668)y-17.3x^2-2{,}090.5x-4{,}008}{y^2}.$

For what value of y does this function have its maximum?

The differential is

$$(334x+668)y^2\,dy$$
$$\underline{-(334x+668)2y^2\,dy+(17.3x^2+2{,}090.5x+4{,}008)2y\,dy}$$

Therefore $(3.34x+668)y=34.6x^2+4{,}181x+8{,}016$

and $y \qquad = \dfrac{34.6x^2+4{,}181x+8{,}016}{3.34x+668}$

This formula gives the values of

$$y \text{ for } x = 0 = {}^{8{,}016}\!\big/_{668} = 12$$
$$y \text{ for } x = 100 \qquad\qquad = 23$$
$$x = 1 \qquad\qquad\quad = 12.2$$

Since for the 12 times interval the interest is $0.12x^2 + 0.84x + 0.72$, for the y times interval

$$\left(0.12x^2+0.84x+0.72\right)\dfrac{144}{y^2}.$$

$$= \dfrac{17.3x^2+121x+103}{y^2}$$

the revenue is

$$\left(\left(334x+\overset{1{,}002}{812}\right)y-2{,}004x-\overset{6{,}012}{4{,}872}\right):y^2$$

Deducting interest of $(121x+103+17.3x^2):y^2$.

Gives ground rent of

$$\frac{(334x+\overset{1,002}{812})y-2,125x-\overset{6,115}{4,975}-17.3x^2}{y^2}.$$

**For what value of y does
this function have its maximum?**

Taking the differential in respect of y and setting it equal to 0 gives:

$$\left(334x+\overset{1,002}{812}\right)y^2\mathrm{d}y+\left(17.3x^2+2,125x+\overset{6,115}{4,975}\right)2y\mathrm{d}y$$

$$-\left(334x+\overset{1,002}{812}\right)y^2\mathrm{d}y$$

$$\overline{\left(334x+\overset{1,002}{812}\right)y=34.6x^2+4,250x+\overset{12,230}{9,950}.}$$

Hence $y=\dfrac{34.6^2+4,250x+\overset{12,230}{9,950}}{334x+\overset{1,002}{812}}$

This gives for $x=10,\ y=\overset{55,910}{}\!/_{4,152}\ =13.5$

For $x=100,\ y=\overset{780,950}{}\!/_{34,212}=22.8$

If $x=60,\ y$ is $\overset{389,510}{}\!/_{20,852}=18.7$

If $x=80,\ y\qquad\qquad=20.8$

If $x=100,\ y\qquad\qquad=22.8$

If $x=120,\ y\qquad\qquad=24.5$

If we include in our calculation not the interest for growth of $6x+18$, but only for the $x+$ 5th year, then

$$y=\frac{34.6x^2+4,180x+12,024}{334x+1,002}.$$

This gives for $x = 0$, $y = 12$
  For $x = 1$, $y = 12.16$

Since for $x = 0$, and for $x = 1$ y is not exactly 12, but very slightly greater, then it is probable that at the beginning of this calculation we should take not the difference between the xth and $x + 1$st years, but that between the $x - 1$st and xth years.

The deviations that arise from this have a very small impact upon the calculated ground rent, and we in any case arrive at a result that is in crass contradiction with prevailing opinion.

Ground rent increases to an extraordinary degree when the tree has become fully formed and is permitted more space, and with this the production costs of strong timber fall in relation to weak timber.

So long as this method is not generally employed, the owner of a wood can make an enormous profit by growing stronger timber.

The increased clearing must be introduced only very gradually, since otherwise storms will blow many trees down.

Beech trees are not susceptible to windthrow. And for firewood the shape of the tree does not matter. While it is necessary for the forming of pines that the wood remains closed up, beech trees can from their early years be thinned according to the propositions developed above – which would have an enormous effect upon the production price of beechwood.

The relation between the girth and height of trees.

In order in the foregoing to follow the proposition of investigating individual factors while abstracting from corresponding factors, I have above set the proportion of girth to height as 1 : h, where h is a constant.

In reality however things are rather different, for the proportion becomes smaller, the taller the tree.

Nagel gives the following proportion:

If the girth is	Then the height is
$\frac{1}{10}$ foot	80 times
$\frac{2}{10}$ foot	79 times
$\frac{3}{10}$ foot	78 times
...	...
1 foot	71 times
2 feet	61 times

If we had expressed this varying proportion as a general formula, this formula would become very complex and calculation very difficult. Application of this formula to *special cases* can easily reveal the impact of varying proportions of length to the timber yield and ground rent.

A more important question to establish the accuracy of our study is however: does this not involve interference with and alteration of those propositions hitherto held to be generally valid?

Before responding to this question, we have to consider another property of trees, which is the gradual loss of the lower branches. Even when thinning is regularly carried out the lower branches, robbed of light, die off, fall to the ground and form loose wood. Over time the yield from such loose wood is not inconsiderable, and according to Nagel covers the losses arising from the reduction in relative height. Or in other words: the growth of the tree, whether young or old, adding the annual fallen wood to the growth of the tree, is always the same for every year, and contributes just as much as when the tree had the proportions of its early years.

Hence the foundation of our investigation remains true: that the growth as well as the girth of individual trees and the entire wood proceeds in an arithmetic progression.

If one adds the fallen timber to that from thinnings, the proportion of these to the yield from clear felling will of course alter, and the former will be more than twice the amount of the latter. This calls for some correction to the value of clear felling and even to the most advantageous rotation period.

But these modifications are not so considerable as to disturb in any way the laws developed above, according to which more vigorous thinning increases ground rent and extends the most advantageous rotation period.

With the increasing clearance around each tree the proportion of girth to height of trees requires clarification in several respects:

1) For trees whose position is isolated the proportion of height to girth is a great deal smaller than for trees growing in a closed stand. Does this proportion change constantly with the increasing interval between trees in a wood, and according to what law?

2) Does the free-standing tree extend its roots downwards to such a great extent and in relation to its girth as the tree in a closed stand, or is a closed stand the condition that compels the deep penetration of the soil?

3) Is the great difference in proportion for trees standing on different kinds of soil not primarily related to the quality of the subsoil?

Epilogue: A Chronicle of Editing the *Isolated State*

Habent sua fata libelli.

Goethe translated Terentianus Maurus's epigram as 'Auch Bücher haben ihr Erlebtes' (books too have their own fate). That is especially true of von Thünen's *Isolated State*.

In *1826* the book was published as *The Isolated State in its Relation to Agriculture and Political Economy or Enquiries on the Effects that have Grain Prices, Fertility of Soil and Taxes on the Agriculture*.

In 1842 followed the Second Edition, now entitled Part I. Only in 1850 came *The Isolated State*, Part II, entitled *The Natural Wage and Its Relation to the Rate of Interest and to Land Rent*.

A detailed chronicle of Part III reads as follows: the manuscript, which is today in the University of Rostock's Thünen Archive, is undated. I owe kind thanks to Angela Hartwig, director of the archive, for allowing me access to it on several occasions.

Von Thünen had worked sporadically on it for many years. For example, in § 17 he notes that the greater part of the preceding § 16 had already been written in 1828. But nothing that went beyond the treatment of forestry in the 'second ring' (*Isolated State*, Part I, § 19; 1826 and 1842) was published before his death in 1850. All that remained was the announcement in the preface to the 1842 Second Edition that a further part would include a more extended study of forestry, together with some other topics.

In 1863 Hermann Schumacher (1827–1904) published the third part of the *Isolated State*. His foreword is dated 24 June 1863 – which would have been von Thünen's 80th birthday – and Schumacher argues that even if the manuscript remained a fragment of a larger work, it was of sufficient importance to be published as it stood.

This third part was republished only twice: in 1875 it was included in the third edition of the *Isolated State*, so that this was the first edition including all three parts. Schumacher was likewise the editor and he prefaced the new edition with a 13-page foreword. In 1966 Walter Braeuer and Eberhardt E. A. Gerhardt issued a reprint of the 1875 Schumacher edition.

All other German editions of the *Isolated State* omit this third part, including the 1986 reprint of the first edition (1826) edited by Horst Claus Recktenwald and the 1990 edition edited by Hermann Lehmann and Lutz Werner. The 1910 edition edited by Heinrich Waentig (often referred to as the Jena edition) also omits the third part; this was reprinted in 1921, 1930, 1966 and 1990, and these editions are the ones today most commonly found in libraries in Germany and elsewhere.

In a thorough book review Jürg Niehans (University of Bern and University of California Santa Cruz) expressed two desiderata:[1] 'A critical edition of Thünen's complete works is overdue. It is gratifying to know that this is

---

1. Book review of Heinz Rieter (ed.), *Studien zur Entwicklung der ökonomischen Theorie*, Bd. 14 (Berlin, 1995). The review was published in *The European Journal of the History of Economic Thought*, Vol. 4, No. 1, Spring 1997, pp. 174–7.

included in the long-range programme of the Thünen Society, though one would have to be an optimist to expect rapid progress. On a more modest scale, an English translation of the last part of *The Isolated State* is also overdue.' Niehans stated this in 1997. Only ten years later one desideratum has now been fulfilled.

Terentianus's epigram cited above runs in full as follows: *Pro captu lectoris habent sua fata libelli* (the fate of books is written in their reader's perception). We trust that the international community of scholars will take up and discuss this translation by Keith Tribe of Part III; and the same goes for the German original, since 'a translation is not the work itself, but a path to the work', as José Ortega y Gasset wrote in his essay 'Miseria y esplendor de la traducción' (Poverty and Grandeur in Translation).

The idea to promote and to sponsor the translation of Part III of the *Isolated State* was born on the occasion of my 70th birthday, on 8 March 2006. Besides thanks to the translator Keith Tribe (University of Sussex), we also thank the editor, Professor Ulrich van Suntum (University of Münster). Professor Heinz Rieter (University of Hamburg) suggested the translator and Professor Heinz Kurz (University of Graz) arranged publication. I thank Angela Ziegler, Librarian at the Thünen Museum in Tellow, for her invaluable assistance in the course of my research.

Reinhard Schwarze,
Hamburg (Honorary Member of the Friends and
Supporters of the Thünen Museum in Tellow/
Mecklenburg)

Glossary

Bushel (*Berliner Scheffel, Berl. Sch.*)
An old way of measuring a unit of volume equivalent to 8 gallons or 56 pounds of rye, which is in metric measurement 40.7 kg of rye.

Cartload (*Fuder, Fuhre*)
An old volume unit holding varying amounts of Lübeck cubic feet, for example, 64 cu. ft. of branchwood or 196 Lübeck cubic feet of firewood.

Clear felling (*Abtriebshieb, Endhieb, Endnutzung*)
Felling of trees at the end of their rotation period. All trees within a compartment are cut down to make way for new planting (reafforestation).

Compartment (*Schlag, Abteilung, Teilfläche, Bestand*)
Area of standing timber. Several terms are to be found in literature: section, district, compartment, cohort, stand, stock, plot of land. The German term *Schlag* was in Thünen's time applied in forestry and agriculture, in our days solely in agriculture. One has to avoid mixing up the terms 'stand' or 'stock' with 'clear felling' or 'final cutting' as the German term *Schlag* is misleading!

Cord (*Faden, Klafter*)
Volume unit of timber measurement equal to 196 Lübeck cubic feet.

Corn (*Korn*)
In Thünen's days the amount of crop was expressed in 'corns'. For example, 'a land yields 10 corns of rye' meant

that there had been a crop of 10 Berlin bushels per 100 square rods. That equals 18.78 decitons per hectare (18,780 kg per 10,000 square metres). Roughly speaking the old English 'ten-fold measure' has the same meaning.

Felling: *see* clear felling.

Lübeck foot
One Lübeck foot equals 0.291024 metre.

Mecklenburg rod
One Mecklenburg rod equals 4.6545 metre

Parisian line
One Parisian line equals 2.256 millimetre.

Pine (*Kiefer, Föhre, Waldkiefer*)
Scotch pine, scientific name: *Pinus sylvestris Linné*. Not to be mixed up with 'Pine (*Pinus pinea*)' growing in the Mediterranean region.

Rotation period (*Umtriebszeit, Produktionsdauer*)
Period between planting and felling compartments of trees. It starts by planting of the trees and ends when it doesn't pay for the forest-owner to wait any longer to start harvesting all the trees of same cohort. A stand of trees reaches the end of its rotation period when diminishing utility begins. Or, in the words of Samuelson: 'Cut trees down to make way for new trees when they are past their best growth rates.'

Shilling (*Schilling*)
Abbreviation: sh. 48 shillings – sometimes with qualification $N\frac{2}{3}$ (new two-thirds) – equally 1 Thaler $N\frac{2}{3}$ in old Mecklenburg currency: *see also* Thaler.

Stand: *see* compartment

Stock: *see* compartment

Square rod
One square rod equals 21.66437 square metres.

Thaler

Abbreviation: Th. A Thaler is equivalent to 48 shillings. Sometimes written with qualifications: Thaler $N\frac{2}{3}$ (new two-thirds). It was the basic unit of Mecklenburg currency between 1789 and 1848.

Thinning (*Durchforstung*)

Regular fellings within a rotation period to improve growing conditions for the remaining trees. A pine forest can be thinned by felling all trees that are not pines, by felling stunted or weak pines, or by felling those which are crowding champion pines. Champion trees are those selected to be sold for the highest attainable price after felling. Such trees are expected to have at the conclusion of their rotation period the best trunks, phenotype, soundness of wood, volume and financial yield.

Waldmorgen

The only term in our book without translation into English. Thünen defines it as follows: 'The area necessary for the realisation of an annual growth of 100 cubic feet I call for the sake of brevity a *Waldmorgen*. The size of the *Waldmorgen* is here, where 100 square rods yield 76.8 cubic feet (of wood), 130 square rods.' That equals 2,816 square metres.

Reinhard Schwarze
Hamburg

References

Gerhard Stinglwagner, Ilse Haseder and Reinhold Erlbeck (2005) *Das Kosmos Wald- und Forstlexikon*, 3rd edn. Stuttgart.

von Thünen, Johann Heinrich (1990) *Der isolierte Staat in Beziehung auf Landwirtschaft und Nationalökonomie*, ed. Hermann Lehmann and Lutz Werner, Berlin.

Warkotsch, Walter (1990) *Forstliches Wörterbuch Deutsch-Englisch*. Remagen-Oberwinter, 2001.